Abrupt Climate Change
Inevitable Surprises

Committee on Abrupt Climate Change
Ocean Studies Board
Polar Research Board
Board on Atmospheric Sciences and Climate
Division on Earth and Life Studies
National Research Council

NATIONAL ACADEMY PRESS
Washington, D.C.

NATIONAL ACADEMY PRESS • 2101 Constitution Ave., N.W. • Washington, DC 20418

NOTICE: The project that is the subject of this report was approved by the Governing Board of the National Research Council, whose members are drawn from the councils of the National Academy of Sciences, the National Academy of Engineering, and the Institute of Medicine. The members of the committee responsible for the report were chosen for their special competences and with regard for appropriate balance.

This study was supported by Contract/Grant 50-DKNA-7-90052 between the National Academy of Sciences, NOAA's United States Global Change Research Program, and the National Aeronautics and Space Administration. Additional funds for the impacts workshop were provided by the Yale National Bureau of Economic Research Program on Environmental Economics. Any opinions, findings, conclusions, or recommendations expressed in this publication are those of the author(s) and do not necessarily reflect the views of the organizations or agencies that provided support for the project.

Library of Congress Cataloging-in-Publication Data

Abrupt climate change : inevitable surprises / Committee on Abrupt Climate Change, Ocean Studies Board, Polar Research Board, Board on Atmospheric Sciences and Climate, Division on Earth and Life Studies, National Research Council.
    p. cm.
Includes bibliographical references and index.
  ISBN 0-309-07434-7
  I. National Research Council (U.S.). Committee on Abrupt Climate Change.

2002002017

Cover: This Spaceborne Imaging Radar-C/X-Band Synthetic Aperture Radar (SIR-C/X-SAR) image provided by NASA JPL and acquired from the Space Shuttle Endeavour, STS-59 (April 11, 1994), shows part of the vast Namib Sand Sea on the west coast of southern Africa, just northeast of the city of Luderitz, Namibia. This region receives only a few centimeters of rain per year. The colors are assigned to different radar frequencies. The magenta areas in the image are fields of sand dunes, and the orange area along the bottom is the surface of the South Atlantic Ocean. The bright green features in the upper right are rocky hills protruding through the sand sea. Because this radar penetrates through the sand, it can reveal sub-surface features such as former lakes, rivers, and drainage channels that have long since dried up as the climate changed. SIR-C/X-SAR is a joint US-German-Italian project that captures sophisticated images of Earth that are useful to scientists from many disciplines. (NASA JPL image).

Additional copies of this report are available from the National Academy Press, 2101 Constitution Avenue, N.W., Lockbox 285, Washington, DC 20055; 800-624-6242 or 202-334-3313 (in the Washington metropolitan area); Internet: http://www.nap.edu.

THE NATIONAL ACADEMIES

National Academy of Sciences
National Academy of Engineering
Institute of Medicine
National Research Council

The **National Academy of Sciences** is a private, nonprofit, self-perpetuating society of distinguished scholars engaged in scientific and engineering research, dedicated to the furtherance of science and technology and to their use for the general welfare. Upon the authority of the charter granted to it by the Congress in 1863, the Academy has a mandate that requires it to advise the federal government on scientific and technical matters. Dr. Bruce M. Alberts is president of the National Academy of Sciences.

The **National Academy of Engineering** was established in 1964, under the charter of the National Academy of Sciences, as a parallel organization of outstanding engineers. It is autonomous in its administration and in the selection of its members, sharing with the National Academy of Sciences the responsibility for advising the federal government. The National Academy of Engineering also sponsors engineering programs aimed at meeting national needs, encourages education and research, and recognizes the superior achievements of engineers. Dr. Wm. A. Wulf is president of the National Academy of Engineering.

The **Institute of Medicine** was established in 1970 by the National Academy of Sciences to secure the services of eminent members of appropriate professions in the examination of policy matters pertaining to the health of the public. The Institute acts under the responsibility given to the National Academy of Sciences by its congressional charter to be an adviser to the federal government and, upon its own initiative, to identify issues of medical care, research, and education. Dr. Kenneth I. Shine is president of the Institute of Medicine.

The **National Research Council** was organized by the National Academy of Sciences in 1916 to associate the broad community of science and technology with the Academy's purposes of furthering knowledge and advising the federal government. Functioning in accordance with general policies determined by the Academy, the Council has become the principal operating agency of both the National Academy of Sciences and the National Academy of Engineering in providing services to the government, the public, and the scientific and engineering communities. The Council is administered jointly by both Academies and the Institute of Medicine. Dr. Bruce M. Alberts and Dr. Wm. A. Wulf are chairman and vice chairman, respectively, of the National Research Council.

Preface

arge, abrupt climate changes have repeatedly affected much or all of the earth, locally reaching as much as 10°C change in 10 years. Available evidence suggests that abrupt climate changes are not only possible but likely in the future, potentially with large impacts on ecosystems and societies.

This report is an attempt to describe what is known about abrupt climate changes and their impacts, based on paleoclimate proxies, historical observations, and modeling. The report does not focus on large, abrupt causes—nuclear wars or giant meteorite impacts—but rather on the surprising new findings that abrupt climate change can occur when gradual causes push the earth system across a threshold. Just as the slowly increasing pressure of a finger eventually flips a switch and turns on a light, the slow effects of drifting continents or wobbling orbits or changing atmospheric composition may "switch" the climate to a new state. And, just as a moving hand is more likely than a stationary one to encounter and flip a switch, faster earth-system changes—whether natural or human-caused—are likely to increase the probability of encountering a threshold that triggers a still faster climate shift.

We do not yet understand abrupt climate changes well enough to predict them. The models used to project future climate changes and their impacts are not especially good at simulating the size, speed, and extent of the past changes, casting uncertainties on assessments of potential future changes. Thus, it is likely that climate surprises await us.

When orbital wiggles and rising greenhouse gases warmed the earth from the last ice age, proxy records show that smooth changes were interspersed with abrupt coolings and warmings, wettings and dryings. By analogy, the expected future warming may come smoothly, but may come with jumps, short-lived or local coolings, floods or droughts, and other unexpected changes. Societies and ecosystems have an easier time dealing with slower or better-anticipated changes, so the abruptness and unpredictability of the possible changes may be disquieting.

This report considers patterns, magnitudes, mechanisms, and impacts of abrupt climate changes, possible implications for the future, and critical knowledge gaps. The potentially large impacts and prediction difficulties focus special attention on increasing the adaptability and resiliency of societies and ecosystems. The committee notes that there is no need to be fatalistic; human and natural systems have survived many abrupt changes in the past, and will continue to do so. Nonetheless, future dislocations can be minimized by taking steps to face the potential for abrupt climate change. The committee believes that increased knowledge is the best way to improve the effectiveness of response, and thus that research on abrupt climate change can help reduce vulnerabilities and increase adaptive capabilities.

I would like to thank the US Global Change Research Program and staff at the many agencies who are a part of USGCRP, for funding and participating in this study process. Thanks also to the Yale/NBER Program on International Environmental Economics for additional funding, and to committee members Bill Nordhaus and Dorothy Peteet for organizing and conducting the Impacts Workshop. The numerous participants at our workshops, the reviewers, and many other colleagues contributed valuable insights and encouragement. It has been my privilege to work with Study Director Alexandra Isern (now with the National Science Foundation), Polar Research Board Director Chris Elfring, Research Associate John Dandelski, and with Senior Project Assistants Megan Kelly, Jodi Bachim, and Ann Carlisle, of the National Research Council. I thank the Ocean Studies Board, the Polar Research Board, and the Board on Atmospheric Science and Climate for providing the impetus to do this study and oversight throughout the process.

I would especially like to extend my deep appreciation to the committee's members for their efforts in creating this report. By exploring new territory and working across disciplines, the committee has opened my eyes to exciting new frontiers. I hope that the readers of this report join us in seeing not peril but opportunity for improved knowledge leading toward a happier and more secure future.

Richard B. Alley, *Chair*
Committee On Abrupt Climate Change

Acknowledgments

The committee would like to express its appreciation to the many people who contributed to this report. In particular, we would like to thank the participants of the two workshops held as part of this study, especially those who gave keynote presentations at the Workshop on Abrupt Climate Change: William Ascher, David Bradford, Grant Branstator, Tony Broccoli, Wallace Broecker, Mark Cane, Bob Dickson, Isaac Held, Sylvie Joussaume, John Kutzbach, Jean Lynch-Stieglitz, Peter Schlosser, Jeff Severinghaus, Karen Smoyer, and Gary Yohe. These talks helped set the stage for fruitful discussions in the breakout sessions that followed. The steering committee is also grateful to Doug Martinson for stepping in at the last minute to lead one of the breakout discussions. In addition, the committee is grateful to the participants of the Workshop on the Economic and Ecological Impacts of Abrupt Climate Change for their insights and written reports, parts of which have been incorporated into this report: Craig Allen, Edward Cook, Peter Daszak, Mark Dyurgerov, David Inouye, Klaus Keller, George Kling, Walter Koenig, Carl Leopold, Thomas Lowell, Robert Mendelsohn, John Reilly, Joel Smith, Thomas Swetnam, Richard Tol, Ferenc Toth, Harvey Weiss, John Weyant, and Gary Yohe. The committee is also grateful to a number of people who provided important discussion and/or material for this report, including Vic Baker, Katherine Hirschboeck, and James Knox.

The committee would like to acknowledge the generous financial support provided by the U.S. Global Change Research Program, which requested this study, and the Yale National Bureau of Economic Research on International Environmental Economics, which provided funds for the impacts workshop.

This report has been reviewed in draft form by individuals chosen for their diverse perspectives and technical expertise, in accordance with procedures approved by the National Research Council's Report Review Committee. The purpose of this independent review is to provide candid and critical comments that will assist the institution in making its published report as sound as possible and to ensure that the report meets institutional standards for objectivity, evidence, and responsiveness to the study charge. The review comments and draft manuscript remain confidential to protect the integrity of the deliberative process. We wish to thank the following individuals for their participation in the review of this report:

James Kennett, University of California, Santa Barbara
Mahlon C. Kennicutt, Texas A&M University, College Station
Roger Lukas, University of Hawaii, Manoa
Robie Macdonald, Fisheries and Oceans Canada, British Columbia
Vera Markgraf, University of Colorado, Boulder
Stephen Rayner, Columbia University, New York
Jeffrey Severinghaus, Scripps Institution of Oceanography, La Jolla
Andrew Weaver, University of Victoria, British Columbia, Canada
Gary Yohe, Wesleyan University, Middletown, Connecticut

Although the reviewers listed above have provided many constructive comments and suggestions, they were not asked to endorse the conclusions or recommendations nor did they see the final draft of the report before its release. The review of this report was overseen by Robert Knox (Scripps Institution of Oceanography) and Stephen Berry (University of Chicago), appointed by the National Research Council, who were responsible for making certain that an independent examination of this report was carried out in accordance with institutional procedures and that all review comments were carefully considered. Responsibility for the final content of this report rests entirely with the authoring committee and the institution.

Contents

Abrupt Climate Change

Executive Summary

ecent scientific evidence shows that major and widespread climate changes have occurred with startling speed. For example, roughly half the north Atlantic warming since the last ice age was achieved in only a decade, and it was accompanied by significant climatic changes across most of the globe. Similar events, including local warmings as large as 16°C, occurred repeatedly during the slide into and climb out of the last ice age. Human civilizations arose after those extreme, global ice-age climate jumps. Severe droughts and other regional climate events during the current warm period have shown similar tendencies of abrupt onset and great persistence, often with adverse effects on societies.

Abrupt climate changes were especially common when the climate system was being forced to change most rapidly. Thus, greenhouse warming and other human alterations of the earth system may increase the possibility of large, abrupt, and unwelcome regional or global climatic events. The abrupt changes of the past are not fully explained yet, and climate models typically underestimate the size, speed, and extent of those changes. Hence, future abrupt changes cannot be predicted with confidence, and climate surprises are to be expected.

The new paradigm of an abruptly changing climatic system has been well established by research over the last decade, but this new thinking is little known and scarcely appreciated in the wider community of natural and social scientists and policy-makers. At present, there is no plan for improving our understanding of the issue, no research priorities have been identified, and no policy-making body is addressing the many concerns raised by the potential for abrupt climate change. Given these gaps, the US Global Change Research Program asked the National Research Council to establish the Committee on Abrupt Climate Change and charged the group to describe the current state of knowledge in the field and recommend ways to fill in the knowledge gaps.

It is important not to be fatalistic about the threats posed by abrupt climate change. Societies have faced both gradual and abrupt climate changes for millennia and have learned to adapt through various mechanisms, such as moving indoors, developing irrigation for crops, and migrating away from inhospitable regions. Nevertheless, because climate change will likely continue in the coming decades, denying the likelihood or downplaying the relevance of past abrupt events could be costly. Societies can take steps to face the potential for abrupt climate change. The committee believes that increased knowledge is the best way to improve the effectiveness of response, and thus that research into the causes, patterns, and likelihood of abrupt climate change can help reduce vulnerabilities and increase our adaptive capabilities. The committee's research recommendations fall into two broad categories: (1) implementation of targeted research to expand instrumental and paleoclimatic observations and (2) implementation of modeling and associated analysis of abrupt climate change and its potential ecological, economic, and social impacts. What follows is a summary of recommended research activities; more detail is presented in the chapters, particularly in Chapter 6.

IMPROVE THE FUNDAMENTAL KNOWLEDGE BASE RELATED TO ABRUPT CLIMATE CHANGE

Recommendation 1. Research programs should be initiated to collect data to improve understanding of thresholds and nonlinearities in geophysical, ecological, and economic systems. Geophysical efforts should focus especially on modes of coupled atmosphere-ocean behavior, oceanic deepwater processes, hydrology, and ice. Economic and ecological research should focus on understanding nonmarket and environmental issues, initiation of a comprehensive land-use census, and development of integrated economic and ecological data sets. These data will enhance understanding of abrupt climate change impacts and will aid development of adaptation strategies.

Physical, ecological, and human systems are imperfectly understood, complex, nonlinear, and dynamic. Current changes in climate are producing conditions in these systems that are outside the range of recent historical experience and observation, and it is unclear how the systems will interact with and react to the coming climatic changes. Our ability to adapt to or mitigate the effects of climate change will be improved if we can recognize climate-related changes quickly. This will require improved monitoring of climatic, ecological, and socioeconomic systems. Many of the needed data sets overlap with those used to study gradual climate change.

To increase understanding of abrupt climate change, research should be directed toward aspects of the climate system that are believed to have participated in past abrupt changes or that are likely to exhibit abrupt and persistent changes when thresholds in the climate system are crossed. Key research areas for increasing our understanding of abrupt climate change include:

- oceanic circulation, especially related to deepwater formation;
- sea-ice transport and processes, particularly where they interact with deepwater formation;
- land-ice behavior, including conditions beneath ice sheets;
- the hydrological cycle, including storage, runoff, and permafrost changes; and
- modes of atmospheric behavior and how they change over time.

In the ecological and human sphere, data collection should target sectors where the impacts of abrupt climate change are likely to be largest or where knowledge of ongoing changes will be especially useful in understanding impacts and developing response alternatives. Data collection should include a comprehensive land-use census that monitors fragmentation of ecosystems, tracking of wildlife diseases, and conditions related to forest fires, as well as improved seasonal and long-term climate forecasts, and sustained study of oceanic regimes of intense biological activity, particularly near the coasts. In the social arena, priority should be given to development of environmental and nonmarket accounts, and analyses of possible threshold crossings.

IMPROVE MODELING FOCUSED ON ABRUPT CLIMATE CHANGE

> **Recommendation 2.** New modeling efforts that integrate geophysical, ecological, and social-science analyses should be developed to focus on investigating abrupt climate changes. In addition, new mechanisms that can cause abrupt climate change should be investigated, especially those operating during warm climatic intervals. Understanding of such mechanisms should be improved by developing and applying a hierarchy of models, from theory and conceptual models through models of intermediate complexity, to high-resolution models of components of the climate system, to fully coupled earth-system models. Model-data comparisons should be enhanced by improving the ability of models to simulate changes in quantities such as isotopic ratios that record past climatic conditions. Modeling should be used to generate scenarios of abrupt climate change with high spatial and temporal resolution for assessing impacts and testing possible adaptations. Enhanced, dedicated computational resources will be required for such modeling.

Developing theoretical and empirical models to understand abrupt climate changes and the interaction of such changes with ecological and economic systems is a high priority. Modeling is essential for collaborative research between physical, ecological, and social scientists, and much more effort is needed to develop accurate models that produce a useful understanding of abrupt climate processes. Model analyses help to focus research on possible causes of abrupt climate change, such as human activities; on key areas where climatic thresholds might be crossed; and on fundamental uncertainties in climate-system dynamics. To date, most analyses have considered only gradual climate change; given the accumulating evidence of past abrupt climate change and of its capacity to affect human societies, more attention should be focused on scenarios involving abrupt change.

Climate models that are used to test leading hypotheses for abrupt climate change, such as altered deep-ocean circulation, can only partially simulate the size, speed, and extent of the large climatic changes that have occurred. The failure to explain the climate record fully suggests either that the proposed mechanisms being used to drive these models are incomplete or that the models are not as sensitive to abrupt climate change as is the natural environment. It is also of concern that existing models do not accurately simulate warm climates of the past.

A comprehensive modeling strategy designed to address abrupt climate change should include vigorous use of a hierarchy of models, from theory and conceptual models through models of intermediate complexity, to high-resolution models of components of the climate system, to fully coupled earth-system models. The simpler models are well suited for use in developing new hypotheses for abrupt climate change and should focus on warmer climate, because warming is likely. Because reorganizations of the thermohaline circulation have never been demonstrated in climate models employing high-resolution ocean components, improving the spatial resolution in climate models assumes high priority. Complex models should be used to produce geographically resolved (to about 1° of latitude by 1° of longitude), short-time (annual or seasonal) sensitivity experiments and scenarios of possible abrupt climatic changes.

Long integrations of fully coupled models under various forcings for the past, present, and future are needed to evaluate the models, assess possibilities of future abrupt changes, and provide scenarios of those future changes. The scenarios can be combined with integrated-assessment economic models to improve understanding of the costs for alternative adaptive approaches to climate change with attention to the effects of rising greenhouse-gas concentrations and nonclimatic factors, such as land-use changes and urbanization. Model-data comparisons are needed to assess the quality of model predictions. It is important to note that the multiple long integrations of enhanced, fully coupled earth-system models required for this research are not possible with the computer resources available today, and thus these resources should be enhanced.

IMPROVE PALEOCLIMATIC DATA RELATED
TO ABRUPT CLIMATE CHANGE

Recommendation 3. The quantity of paleoclimatic data on abrupt change and ecological responses should be enhanced, with special emphasis on:

• Selected coordinated projects to produce especially robust, multi-parameter, high-resolution histories of climate change and ecological response.
• Better geographic coverage and higher temporal resolution.
• Additional proxies, including those that focus on water (e.g., droughts, floods, etc.).

- Multidisciplinary studies of selected abrupt climate changes.

The current scientific emphasis on abrupt climate change was motivated by strong evidence in proxy records that showed extreme climatic changes in the past, sometimes occurring within periods of fewer than 10 years. Paleoclimatic records provide important information related to changes in many environmental variables. However, not all proxy archives provide equally high confidence for estimating past climatic conditions, such as temperature and precipitation, and for determining when and how rapidly changes occurred.

Confidence can be improved by encouraging coordinated, multi-parameter, multi-investigator study of selected archives that have seasonal to decadal time accuracy and resolution, substantial duplication of measurements to demonstrate reproducibility, and extensive calibration of the relation between climate and sedimentary characteristics. As one example, in the ice-core projects from central Greenland, duplication of the measurements by independent, international teams provides exceptional confidence in most data and reveals which data sets do not warrant confidence. Sampling at very high time resolution to produce data sets complementary to those of other investigators gives an exceptionally clear picture of past climate. Such projects require more funding and effort than are typical of paleoclimatic research, but they provide an essential reference standard of abrupt climate change to which other records can be compared. A difficulty is that this reference standard is from one place in high northern latitudes and is inappropriate for study of much of the climate system.

Not all paleoclimatic records can be studied in the same detail as those from Greenland, but generation of at least a few similar highly resolved (preferably annually or subannually) reference standards including a North Atlantic marine record comparable with Greenland records, would be of great value. The ultimate goal is to develop a global network of records with at least decadal resolution. Terrestrial and marine records of climate change and ecological response from the regions of the western Pacific warm pool (the warmest part of the global climate system) and the Southern Ocean and Antarctic continent (the southern cold pole of the climate system) are among the most critical targets for future paleoclimate research, including generation of reference standards.

Abrupt climate change is likely to influence water availability and therefore is of great concern for economic and ecological systems. Focus on measures of precipitation, evaporation, and the quantitative difference between them is particularly important. Freshwater balance is also important in controlling water density and thus the thermohaline circulation of the oceans; reconstructions of water-mass density in polar and subpolar regions are central. New methods for investigating past changes in the hydrological cycle are important, as are additional studies of the relation between a range of climatic changes and the signals they leave in sedimentary archives.

Global maps of past climates, with high resolution in time and space and spanning long intervals, would be of great use to the climate community. However, such maps are unlikely to be available soon. The traditional alternative of reconstructing climate for selected moments, or "time-slices," fails to capture the short-lived anomalies of abrupt climate changes. Instead, mapping efforts are needed and should focus on the patterns of selected abrupt climatic changes in time and space and on their resulting effects. Additional emphasis on annually resolved records of the last 2,000 years will help to place the warming and associated changes of the last 100 years in context.

IMPROVE STATISTICAL APPROACHES

Recommendation 4. Current practices in the development and use of statistics related to climate and climate-related variables generally assume a simple, unchanging distribution of outcomes. This assumption leads to serious underestimation of the likelihood of extreme events. The conceptual basis and the application of climatic statistics should be re-examined with an eye to providing realistic estimates of the likelihood of extreme events.

Many societal decisions are based on assumptions about the distribution of extreme weather-related events. Large capital projects, for instance, often have embedded safety margins that are derived from data and assumptions about the frequency distribution of extreme events. Many major decisions are based on statistical calculations that are appropriate for stationary climates, such as in the use of "30-year normals," for deriving climate data for individual locations.

On the whole, those assumptions are reasonable, if imperfect, rules of thumb to use when the variability of weather is small and climate is stationary. If climate follows normal distributions with known and constant means and standard deviations, businesses and governments can use current practices. However, in light of recent findings related to nonstationary and often highly skewed climate-related variables, current practices can be misleading and result in costly errors.

The potential for abrupt climate change and the existence of thresholds for its effects require revisions of our statistical estimates and practices.

INVESTIGATE "NO-REGRETS" STRATEGIES
TO REDUCE VULNERABILITY

Recommendation 5. Research should be undertaken to identify "no-regrets" measures to reduce vulnerabilities and increase adaptive capacity at little or no cost. No-regrets measures may include low-cost steps to: slow climate change; improve climate forecasting; slow biodiversity loss; improve water, land, and air quality; and develop institutions that are more robust to major disruptions. Technological changes may increase the adaptability and resiliency of market and ecological systems faced by the prospect of damaging abrupt climate change. Research is particularly needed to assist poor countries, which lack both scientific resources and economic infrastructure to reduce their vulnerabilities to potential abrupt climate changes.

Social and ecological systems have long dealt with climate variability by taking steps to reduce vulnerability to its effects. The rapidity of abrupt climate change makes adaptation more difficult. By moving research and policy in directions that will increase the adaptability of economic and ecological systems, it might be possible to reduce vulnerability and increase adaptation at little or no cost. Many current policies and practices are likely to be inadequate in a world of rapid and unforeseen climatic changes. Improving these policies will be beneficial even if abrupt climate change turns out to fit a best-case, rather than a worst-case, scenario. Societies will have "no regrets" about the new policies, because they will be good policies regardless of the magnitude of environmental change. For example, the phaseout of chloroflourocarbons and replacement by gases with shorter atmospheric lifetimes have reduced the US contribution to global warming while at the same time reducing future health risks posed by ozone depletion.

In land-use and coastal planning, managers should consider the effects on ecosystem services that could result from interaction of abrupt climate changes with changes caused by people. Scientists and government organizations at various levels may be used to develop and implement regulations and policies that reduce environmental degradation of water, air, and biota. Conservation measures related to land and watersheds might be put into place to reduce the rate of biotic invasions, with management strategies used to limit the spread of invasions. The potential economic and ecological costs of disease emerging from abrupt climate change should be assessed.

A promising option is to improve institutions to allow societies to withstand the greater risks associated with abrupt changes in climate. For example, water systems are likely to be stressed by abrupt climate change; to manage scarce water, it might prove beneficial to seek more flexible ways to allocate water, such as through use of water markets. Another example of a "no-regrets" strategy is insurance against the financial impacts of fires, floods, storms, and hurricanes. Through the development of new instruments, such as weather derivatives and catastrophe bonds, markets might better accommodate extreme events such as the effects of abrupt climate change. It will be important to investigate the development of better instruments to spread large losses that result from extreme events, priced realistically to reflect the risks but not to encourage excessive risk taking.

Because of the strength of existing infrastructure and institutions, the United States and other wealthy nations are likely to cope with the effects of abrupt climate change more easily than poorer countries. That does not mean that developed countries can remain isolated from the rest of the world, however. With growing globalization, adverse impacts—although likely to vary from region to region because exposure and sensitivity will vary—are likely to spill across national boundaries, through human and biotic migration, economic shocks, and political aftershocks. Thus, even though this report focuses primarily on the United States, the issues are global and it will be important to give attention to the issues faced by poorer countries that are likely to be especially vulnerable to the social and economic impacts of abrupt climate change.

The United States is uniquely positioned to provide both scientific and financial leadership, and to work collaboratively with scientists around the world, to gain better understanding of the global impacts of abrupt climate change as well as reducing the vulnerability and increasing the adaptation in countries that are particularly vulnerable to these changes. Many of the recommendations in this report, although currently aimed at US institutions, would apply throughout the world.

1

Introduction to Abrupt Changes in the Earth's Climate

arge, abrupt climate changes have affected hemispheric to global regions repeatedly, as shown by numerous paleo-climate records (Broecker, 1995, 1997). Changes of up to 16°C and a factor of 2 in precipitation have occurred in some places in periods as short as decades to years (Alley and Clark, 1999; Lang et al., 1999). However, before the 1990s, the dominant view of past climate change emphasized the slow, gradual swings of the ice ages tied to features of the earth's orbit over tens of millennia or the 100-million-year changes occurring with continental drift. But unequivocal geologic evidence pieced together over the last few decades shows that climate can change abruptly, and this has forced a reexamination of climate instability and feedback processes (NRC, 1998). Just as occasional floods punctuate the peace of river towns and occasional earthquakes shake usually quiet regions near active faults, abrupt changes punctuate the sweep of climate history.

The climate system in the past has made large jumps between typical patterns of behavior, as in the mechanical analogy presented in Box 1.1. Especially large and abrupt climate changes have occurred repeatedly over the last 100,000 years during the slide into and climb out of the most recent global ice age. Those changes persisted into the current warm period and probably occurred during previous ice ages (Sarnthein et al., 1994; Broecker, 1995, 1997; Alley and Clark, 1999; Stocker, 2000). Our ability to understand the potential for future abrupt changes in climate is limited by our lack of understanding of the processes that control them. For example, mechanisms proposed to explain abrupt climate shifts do not fully describe the patterns of variability seen in either the paleoclimate or the historical records.

Long-term geological records show that in the past there were different stable states of the climate system from those of today. Differences in these climate states involved the coupled atmosphere, ocean, ice, and biological systems. Some of these old states, such as the proposed "snowball earth" (when most or all of the planet was frozen) (Harland, 1964; Caldeira and Kasting, 1992; Kirschvink, 1992; Hoffman et al., 1998), occurred long ago when geological and astronomic conditions were substantially different from today. However, others, such as the warm-polar "hothouse" pattern, were reached relatively recently, when geological conditions were similar to those of our modern earth (Barron, 1987). Despite the recognition that extreme shifts in the climate system can occur, little information is available on whether transitions between climate states are possible under modern or near-future conditions and whether such transitions would be abrupt.

Understanding how and why climate might change abruptly has important implications. The just-completed US National Assessment (National Assessment Synthesis Team, 2000) emphasizes the possible effects of gradual climate change on societies and ecosystems, and it concludes that effects will probably be larger in the case of faster changes that leave less time for adaptation or that give less warning that might make mitigation possible. Thus, when considering the possible causes and impacts of abrupt climate change, numerous important questions arise, including:

- What caused the large, widespread, abrupt climate changes of the past? Could they recur? Might human activities affect the possibility of recurrence?
- To what extent is abruptness a fundamental characteristic of regional and global climate changes? Might future regional changes be abrupt?
- Could global or regional climates shift into modes different from those observed recently, such as the warm-period modes recorded in geological archives? Might such a shift be abrupt?
- How do societies and ecosystems maintain resilience and adaptability? How might these be enhanced in the face of severe tests posed by abrupt climate changes?

Answers to these and related questions are best guided by a focused re-

Box 1.1
Analogy of Abrupt Climate Change

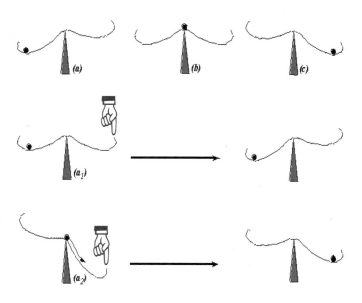

FIGURE 1.1

Abrupt climate change is not unusual, and in fact many simple physical systems exhibit abrupt changes. Here, we illustrate a few basic points using a mechanical analogy.

Imagine a balance consisting of a curved track poised on a fulcrum, as shown above. The track is curved so that there are two "cups" where a ball may rest. A ball is placed on the track and is free to roll until it reaches its point of rest. This system has three *equilibria* denoted (a), (b) and (c) in the top row of the figure. The middle equilibrium (b) is *unstable*: if the ball is displaced ever so slightly to one side or another, the displacement will accelerate until the system is in a state far from its original position. In contrast, if the ball in state (a) or (c) is displaced, the balance will merely rock a bit back and forth, and the ball will roll slightly within its cup until friction restores it to its original equilibrium.

Suppose we push down gently on the right arm of the balance causing a slight tilt, as shown in (a_1). When we let go, the ball will rattle around for a bit as the balance tilts back and forth. Once things settle down, the system will return to its original state of rest, with the ball in the left cup. As noted above, this position is *stable* in the face of small perturbations.

If instead we push the right arm somewhat farther down, as shown in (a_2), the ball will eventually roll over the fulcrum and slide down into the right cup. This is an example of a system passing a *threshold*. When the pressure is relieved, the system does not return to its original state. A temporary influence can have permanent effects; this is what is known as *hysteresis*.

This device illustrates other kinds of behavior that are common in the climate system. The equilibria illustrated in the top row are steady, in that the system sits still without moving. But suppose that the ball in state (a) or (c) is given a gentle push. If the friction is low, the ball in either case will rattle around for a long time, but will remain in its original cup. This illustrates the notion of an unsteady *regime*—the "left cup" regime and the "right cup" regime. A strong enough push at the right time could cause a transition between one regime and the other.

An unusual application of force could cause unexpected behavior. Hit it hard enough, and the device might do something different from anything seen before. For example, the arm of the balance might bang against the table, and the ball could bounce out of the cup and roll away.

Now imagine that you have never seen the device and that it is hidden in a box in a dark room. You have no knowledge of the hand that occasionally sets things in motion, and you are trying to figure out the system's behavior on the basis of some old 78-rpm recordings of the muffled sounds made by the device. Plus, the recordings are badly scratched, so some of what was recorded is lost or garbled beyond recognition. If you can imagine this, you have some appreciation of the difficulties of paleoclimate research and of predicting the results of abrupt changes in the climate system.

search strategy, and this report seeks to identify the key supporting research needs.

DEFINITION OF ABRUPT CLIMATE CHANGE

What defines a climate change as abrupt? Technically, an abrupt climate change occurs when the climate system is forced to cross some threshold, triggering a transition to a new state at a rate determined by the climate system itself and faster than the cause. Chaotic processes in the climate system may allow the cause of such an abrupt climate change to be undetectably small.

To use this definition in a policy setting or public discussion requires some additional context, as is explored at length in Chapter 5, because while many scientists measure time on geological scales, most people are concerned with changes and their potential impacts on societal and ecological time scales. From this point of view, an abrupt change is one that takes place so rapidly and unexpectedly that human or natural systems have difficulty adapting to it. Abrupt changes in climate are most likely to be significant, from a human perspective, if they persist over years or longer, are larger than typical climate variability, and affect sub-continental or larger regions. Change in any measure of climate or its variability can be abrupt, including change in the intensity, duration, or frequency of extreme events. For example, single floods, hurricanes, or volcanic eruptions are important for humans and ecosystems, but their effects generally would not be considered abrupt climate changes unless the climate system is pushed over a threshold into a new state; however, a rapid, persistent change in the number or strength of floods or hurricanes might be an abrupt climate change.

The quintessential abrupt climate change was the end of the Younger Dryas interval about 11,500 years ago, when hemispheric to global climate shifted dramatically, in many regions by about one-third to one-half the difference between ice-age and modern conditions, with much of the change occurring over a few years (Alley, 2000). The changes affected many environmental parameters such as temperature and rainfall (Figure 1.2). Weaker, but still of hemispheric extent, was a short cooling spell 8,200 years ago that lasted for about 200 years (Alley et al., 1997). Although more regionally limited, the apparent change in El Niño behavior toward generally warmer and wetter conditions around 1976 (Nitta and Yamada, 1989; Trenberth, 1990; Graham, 1994) could also be considered an abrupt change. Thus, studies of abrupt climate change overlap with studies of ice

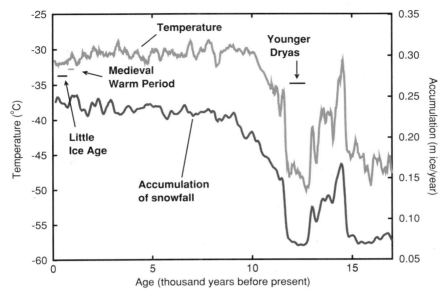

FIGURE 1.2   Climate changes in central Greenland over the last 17,000 years. Reconstructions of temperature and snow accumulation rate (Cuffey and Clow, 1997; Alley, 2000) show a large and rapid shift out of the ice age about 15,000 years ago, an irregular cooling into the Younger Dryas event, and the abrupt shift toward modern values. The 100-year averages shown somewhat obscure the rapidity of the shifts. Most of the warming from the Younger Dryas required about 10 years, with 3 years for the accumulation-rate increase (Figure 2.2). A short-lived cooling of about 6°C occurred about 8,200 years ago (labeled 8ka event), and is shown with higher time resolution in Figure 2.3. Climate changes synchronous with those in Greenland affected much of the world, as shown in Figures 2.1 and 2.3.

ages and other features of deeper time and with studies of decadal-centennial climate modes and variability. In focusing on the tendency of climate to change in fits and starts rather than smoothly, and thus to surprise humans and ecosystems, the study of abrupt climate change is distinct from related branches of climatology.

SCIENTIFIC EVIDENCE, PROCESSES, AND CONSEQUENCES FOR SOCIETY AND ECOSYSTEMS

Abrupt climate change has affected societies. For example, evidence in geologic records suggests that abrupt but persistent droughts caused social disruption for Mayan culture (Hodell et al., 1995; Gill 2000) and that abrupt climate shifts played a role in the collapse of Mesopotamian civilization (Weiss et al., 1993). Recognizing the potential for abrupt changes in climate has constituted a paradigm shift for the research community, but many questions concerning the processes that cause and mediate abrupt climate change remain, including the following.

- What are the patterns of environmental variability associated with abrupt climate change in the tropics and high latitudes?
- What is the role of freshwater cycling in abrupt climate change?
- Will warmer climates influence the occurrence of abrupt climate change?
- Might climate changes occur that are unassociated with a change in external forcing?
- What feedback processes are dominant, and what is their role in causing the persistence of abrupt changes including droughts?

Recent research has shown that human activities are affecting climate, but it is often difficult to separate human-induced changes from those occurring naturally (Intergovernmental Panel on Climate Change, 2001b). The question arises whether anthropogenic influences will trigger abrupt climate change. It is not now possible to answer that question, because the processes that cause abrupt climate change are not sufficiently understood.

There is little doubt that the rate, magnitude, and regional extent of abrupt transitions to different climate states could have far-reaching implications for society and ecosystems. Research has shown that, in response to gradual changes in climate, much of the economic capital stock[1] and some plants and animals may adjust without major disruption. But rapid changes in climate could have major effects, disrupting ecological or economic systems in a manner that prevents their timely replacement, repair, or adaptation.

Ecological systems are particularly vulnerable to abrupt climate change because they are long-lived and relatively immobile. In addition, these systems often have low adaptive capability. Their vulnerability is increased by human activities that alter ecosystems, reducing species abundance and composition and blocking migration. One reason for the vulnerability of eco-

---

[1]Society's capital stock includes both tangible goods such as houses, equipment, and infrastructure as well as intangibles such as patents, education, and ecosystems.

logical and economic systems to the effects of abrupt climate change is that these systems are peculiar to particular locations and adapted to particular climates. Effects are likely to increase when abrupt climate change causes ecological systems to cross thresholds. For example, climatic conditions change greatly over short distances in some areas, and cause similarly steep gradients in vegetation types, allowing even a small climate change to cause dramatic change in vegetation type in a given locale. Rapid climate change will probably result in the redistribution—and possibly in the extinction—of terrestrial and marine species and have major effects on ecosystems worldwide.

Water resources might be greatly affected by abrupt changes in climate. Changes in water supplies could result in increased demands for water, affect agricultural production, and potentially trigger adverse health effects. Those consequences and the economic effects resulting from them provide a strong motivation for enhancing the understanding of physical processes that cause abrupt climate change.

There is increased effort to understand the probability, rate, and magnitude of abrupt climate change. But there is virtually no research on the economic and ecological impacts of abrupt climate change; most research has concentrated on gradual change. It is important to improve knowledge of climate processes that lead to abrupt climate change, and knowledge of the impacts of such changes.

THE CHARGE TO THE COMMITTEE

Because there was no comprehensive review of the science and potential impacts of abrupt climate change and no overall plan for improving the understanding of abrupt climate change, the US Global Change Research Program asked the National Research Council for assistance. In response, the NRC formed the Committee on Abrupt Climate Change, which was charged to:

• describe what is currently known from paleoclimate proxies, historical observations, and numerical modeling about abrupt climate change, including patterns and magnitudes of possible changes, mechanisms, forcing thresholds, and probability of occurrence (at least qualitatively);
• identify the critical knowledge gaps concerning the potential for future abrupt climate changes, including those aspects of change that are of potential importance to society and economies; and

- outline a research strategy to fill the critical knowledge gaps.

A committee of 11 members was appointed from the paleoclimate, climate modeling, and observational climate communities, including those investigating the cryosphere, atmosphere, ocean, and terrestrial systems. To help to focus priorities in abrupt climate change research, representatives of the economic and social science research communities were also part of the committee. Oversight for the committee was provided jointly by the Ocean Studies Board, the Polar Research Board, and the Board on Atmospheric Science and Climate.

To conduct its work, the committee met three times and held two workshops to allow for broad community participation. The first workshop, at Lamont-Doherty Earth Observatory, brought together members of the physical science community as well as those from the social science community who specialize in the societal impacts of climate change. From this workshop, it was decided that to properly address the ecological and societal impacts of abrupt climate change, an additional two-day workshop concentrating on these issues would be needed. This meeting, held in Washington, D.C., was primarily funded by the Yale/NBER Program on International Environmental Economics and provided essential information used to develop the climate impacts portion of this report.

This report is based on the knowledge and experience of the committee's members and input gained from the two workshops. This report discusses the evidence for abrupt climate change (Chapter 2), the processes that can cause these changes (Chapter 3), global warming as a potential trigger for rapid climate change (Chapter 4), and the potential economic and ecological impacts of abrupt climate change (Chapter 5). The final chapter of the report presents the committee's findings and recommendations.

2

Evidence of Abrupt Climate Change

esearchers first became intrigued by abrupt climate change when they discovered striking evidence of large, abrupt, and widespread changes preserved in paleoclimatic archives. Interpretation of such proxy records of climate—for example, using tree rings to judge occurrence of droughts or gas bubbles in ice cores to study the atmosphere at the time the bubbles were trapped—is a well-established science that has grown much in recent years. This chapter summarizes techniques for studying paleoclimate and highlights research results. The chapter concludes with examples of modern climate change and techniques for observing it. Modern climate records include abrupt changes that are smaller and briefer than in paleoclimate records but show that abrupt climate change is not restricted to the distant past.

INTERPRETATION OF PAST CLIMATIC CONDITIONS
FROM PROXY RECORDS

Paleoclimatic interpretation relies ultimately on the use of the present or recent instrumental records as the key to the past. To accomplish this, modern values observed for a given characteristic of the climate system are compared with some record from the past, such as tree-ring thickness or the isotopic composition of water frozen in ice cores (see Plates 1 and 2). Detailed understanding of these records—how the thickness of tree rings changes in recent wet and dry periods—lets scientists draw inferences about the past, and these records come to be considered "proxies," or indicators of the past environment.

The assumption of constancy of the relation between climate and its proxy might require little more to support it than constancy of physical law (for example, the assumption that in the past heat flowed from warm to cold rocks in the same way as today). Other assumptions might involve greater uncertainty (for example, the assumption that under different climatic conditions, marine organisms grew most vigorously during the same season and at the same water depth as in the modern environment). Testing of the underlying assumption that the present is the key to the past relies largely on the consistency of results from a wide array of proxies, particularly those depending on few assumptions. The use of multiple indicators increases the reliability of many paleoclimate reconstructions.

The following pages provide a brief synopsis of paleoclimate proxies (Table 2.1) and age indicators. The description is not exhaustive and is intended only to orient the reader to some of the current paleoclimatic tools available. For more detailed reviews of methods involved in paleoclimatic interpretation see Broecker (1995), Bradley (1999), or Cronin (1999).

Physical paleoclimatic indicators often rely on the fewest assumptions and so can be interpreted most directly. For example, old air extracted from bubbles in ice cores and old water from pore spaces in seabed sediments or continental rocks provide direct indications of past compositions of atmosphere, oceans, and groundwater (see Plate 1). Anomalously cold buried rocks or ice have not finished warming from the ice age and thus provide evidence that conditions were colder in the past. Conditions are also judged from the concentrations of noble gases found dissolved in old groundwaters. Some such records are subject to substantial loss of information through diffusion of the components being analyzed, which limits the ability to interpret older events. Physical indicators include the characteristics of sediments and land features. For example, the presence of sand dunes can indicate past arid conditions, and glacially polished bedrock is an indication of prior glacial conditions.

Isotopic indicators are widely used in paleoclimate science. The subtle differences in behavior between chemically similar atoms having different weights (isotopes) prove to be sensitive indicators of paleoenvironmental conditions. One common application is paleothermometry. The physical and chemical discrimination of atoms of differing isotopic mass increases with decreasing temperature. For example, carbonate shells grow-

TABLE 2.1 Paeloclimatic Proxies

Paleoclimatic Recorder	Climate Variable Recorded	Property Measured
Ice	Atmospheric composition	Trapped bubbles
	Windiness	Dust grain sizes
	Source strength of wind-blown materials	Abundance of pollen, dust, sea salt
	Temperature	Ice isotopic ratios
		Borehole temperatures
		Gas isotopes
		Melt layers
	Snow accumulation rate	Thickness of annual layers
		In-situ radiocarbon
Ocean sediments and corals	Temperature	Species assemblages
		Shell geochemistry
		Alkenone ($U_{37}^{K'}$) thermometry
	Salinity	Shell isotopes after correction for temperature and ice volume
	Ice volume	Isotopic composition of pore waters
		Shell isotopes after correction for temperature and salinity
	pH	Boron isotopes in shells
	Ocean circulation	Cd/Ca in shells
		Carbon-isotopic data
	Corrosiveness/chemistry of ambient waters	Shell dissolution
Lake and bog sediments	Temperature	Species assemblages
		Shell geochemistry
	Atmospheric temperature and soil moisture	Washed- or blown-in materials including pollen and spores
		Macrofossils such as leaves, needles, beetles, midge flies, etc.
	Water balance (precipitation minus evaporation	Species assemblages
		Shell geochemistry
Tree rings	Temperature and/or moisture availability	Ring width or density of trees stressed by cold or drought
	Variations in the isotopic ratio of water related to temperature	Cellulose isotopic ratios
Speleothems/cave formations	Moisture availability	Growth rate of formations
	Isotopic ratios of water related to temperature or precipitation rate	Oxygen isotopic composition
	Overlying vegetation	Carbon-isotopic composition

Table continued on next page

TABLE 2.1 Continued

Paleoclimatic Recorder	Climate Variable Recorded	Property Measured
Terrestrial sediment types/ nature of erosion	Temperature	Glaciers Permafrost
	Snowfall/rainfall	Lakes Sand dunes Glaciers Loess
	Windiness	Loess Sand dunes
	Soil formation rate/moisture availability	Soil profiles Loess
Boreholes	Temperature	Direct measurements
Old groundwater	Temperature	Isotopic and noble gas composition of water
Desert varnish	Moisture availability	Growth rate Chemistry

NOTE: Past climate conditions can be measured only through "proxies," characteristics that give insights about past conditions. For example, gas bubbles trapped in ice can be analyzed to understand the atmosphere at the time the bubbles were trapped. This table lists examples of paleoclimatic proxies, what the proxy measures, and from where the proxy data originated.

ing in water typically favor isotopically heavy oxygen and become isotopically heavier at lower temperatures. Isotopic ratios also are used to estimate the concentration of a chemical. When a chemical is common in the environment, a "favored" isotope will be used; shortage of a chemical leads to greater use of a less favored isotope. Marine photosynthesis increasingly favors the light isotope of carbon as carbon dioxide becomes more abundant, and this allows estimation of changes in carbon dioxide concentration from the isotopic composition of organic matter in oceanic sediments. Similarly, the growth of ice sheets removes isotopically light water (ordinary water) from the ocean, increasing the use of isotopically heavy oxygen from water in carbonate shells, which then provide information on the size of ice sheets over time. Stable isotopic values in organic matter also provide important information on photosynthetic pathways and so can afford insight into the photosynthesizing organisms that were dominant at a given location in the past.

Many chemical proxies of environmental change act like isotopic ratios in the measurement of availability of a species. For example, if decreased rainfall increases the concentration of magnesium or strontium ions in lake water, they will become more common in calcium-carbonate shells that grow in that water. However, warming can also allow increased incorporation of substitute ions in shells. Such nonuniqueness can usually be resolved through use of multiple indicators. Other chemical indicators are allied to biological processes. For example, some species of marine diatoms incorporate stiffer molecules in their cell walls to offset the softening effects of higher temperature, and these molecules are resistant to changes after the diatoms die. The fraction of stiffer molecules in sediments yields an estimate of past temperatures. This analytic technique, known as alkenone paleothermometry, is increasingly used to learn about paleotemperatures in the marine environment.

Biological indicators of environmental conditions typically involve the presence or absence of indicator species or assemblages of species. For example, the existence of an old rooted tree stump shows that the climate was warm and wet enough for trees, and the type of wood indicates how warm and wet the climate was; if that tree stump is in a region where trees do not grow today, the climate change is clear. In ocean and lake sediments, the microfossil species present can indicate the temperature, salinity, and nutrient concentration of the water column when they were deposited. Pollen and macrofossils preserved in sediments are important records of variability in the terrestrial environment (see Plate 3). The presence of specific organic compounds called biomarkers in sediments can reveal what species were present, how abundant they were, and other information.

The complicated nature of paleoclimatic interpretation can be seen when proxies are viewed in a practical example. During ice ages, the oceans were colder, but the water in them was also isotopically heavier because light water was removed and used in growing ice sheets. Shells that grew in water during ice age intervals contain heavier isotopes owing to cooling and changes in the isotopic composition of ocean waters. The change in ocean isotopic composition can be estimated independently from the composition of pore waters in sediments, whereas the change in temperature can be estimated from both the abundance of cold- or warm-loving shells in sediment and the abundance of stiff diatom cell-wall molecules in sediments. Concentrations of non-carbonate ions substituted into calcium carbonate shells provide further information. Because there is redundancy in the available data, reliable results can be obtained.

Any paleoclimatic record requires age estimates, and many techniques are used to obtain them. Annual layers in trees, in sediments of some lakes and shallow marine basins, in corals, and in some ice cores allow high-resolution dating for tens of thousands of years, or longer in exceptional cases. Various radiometric techniques are also used. Dates for the last 50,000 years are most commonly obtained by using radiocarbon (^{14}C). Changes in production of radiocarbon by cosmic rays have occurred over time, but their effects are now calibrated by using annual-layer counts or other radiometric techniques, such as the use of radioactive intermediates generated during the decay of uranium and thorium and also through the potassium-argon system. Other techniques rely on measurement of accumulated damage to mineral grains, rocks, or chemicals; this permits dating on the basis of cosmogenic exposure ages, thermoluminescence, obsidian hydration, fission tracks, amino-acid racemization, and so on. Numerous techniques allow correlation of samples and assignment of ages from well-dated to initially less well-dated records. Such techniques include the identification of chemically "fingerprinted" fallout from particular volcanic eruptions, of changes in the composition of atmospheric gases trapped in ice cores, and of changes in cosmogenic isotope production or rock magnetization linked to changes in the earth's magnetic field.

THE YOUNGER DRYAS AS AN EXAMPLE OF ABRUPT CLIMATE CHANGE

Sedimentary records reveal numerous large, widespread abrupt climate changes over the last 100,000 years and beyond. The best known of them is the Younger Dryas cold interval. The Younger Dryas was a nearly global event that began about 12,800 years ago when there was an interruption in the gradual warming trend that followed the last ice age. The Younger Dryas event ended abruptly about 11,600 years ago (Figures 2.1 and 2.2). Because the Younger Dryas can be tracked quite clearly in geologic records and has received extensive study, a rather detailed summary of the evidence is given here, followed by briefer reviews of other abrupt climate changes. We then target Holocene[1] abrupt climate events as examples of substantial changes that have taken place when physical conditions on the earth were more similar to today. Understanding the causes of both types of abrupt

---

[1]The Holocene is the most recent 11,000 years since the last major glacial epoch or "ice age."

climate change is essential for assessing the importance of their role in our climate future.

Ice Core Evidence of the Younger Dryas

The Younger Dryas cold reversal is especially prominent in ice-core records from Greenland, but it is also observed in ice cores from other locations. The ice-core records provide a unique perspective that demonstrates the synchronous nature of the large, widespread changes observed.

Annual-layer counting in Greenland ice cores allows determination of the age, duration, and rapidity of change of the Younger Dryas event with dating errors of about one percent (Alley et al., 1993; Meese et al., 1997). Annual-layer thicknesses corrected for the effects of ice flow give the history of snow accumulation rate in Greenland (Alley et al., 1993). Concentrations of wind-blown materials—such as dust (which in central Greenland has characteristics showing its origin in central Asia [Biscaye et al., 1997]) and sea salt—reveal changes in atmospheric concentrations of these particles (Mayewski et al., 1997) after correction for variations in dilution caused by changing snow accumulation rate (Alley et al., 1995a). Gases trapped in bubbles reveal past atmospheric composition. Methane is of special interest because it probably records the global area of wetlands. Furthermore, differences between methane concentrations observed in Greenland ice cores and those from Antarctica allow inference of changes in the wetland areas in the tropics and high latitudes (Chappellaz et al., 1997; Brook et al., 1999).

The combination of the isotopic record of water making up the Greenland ice (see Plate 2; Figure 1.2) (Johnsen et al., 1997; Grootes and Stuiver, 1997) and the physical temperature of the ice (Cuffey et al., 1994, 1995; Johnsen et al., 1995) yields estimates of past temperatures in central Greenland, which can be checked by using two additional thermometers based on the thermal fractionation of gas isotopes after abrupt temperature changes (Severinghaus et al., 1998). Ice-core records from Greenland thus provide high-resolution reconstructions of local environmental conditions in Greenland (temperature and snow accumulation rate), conditions well beyond Greenland (wind-blown materials including sea salt and Asian dust), and even some global conditions (wetland area inferred from methane), all on a common time scale (Figures 2.1, 2.2, and 2.3).

A review of available Greenland ice-core data is given by Alley (2000). The data were collected by two international teams of investigators from multiple laboratories. The duplication shows the high reliability of the

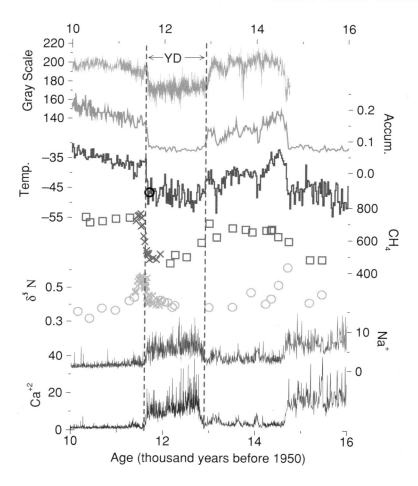

FIGURE 2.1 The Younger Dryas (YD) climate event, as recorded in an ice core from central Greenland and a sediment core from offshore Venezuela. The upper-most curve is the gray-scale (light or dark appearance) of the Cariaco Basin core, and probably records changes in windiness and rainfall (Hughen et al., 1998). The other curves are from the GISP2, Greenland ice core. The rate of snow accumulation and the temperature in central Greenland were calculated by Cuffey and Clow (1997), using the layer-thickness data from Alley et al. (1993) and the ice-isotopic ratios from Grootes and Stuiver (1997), respectively. The independent Severinghaus et al. (1998) temperature estimate is shown by the circle near the end of the Younger Dryas. Methane data are from Brook et al. (1996) (squares) and Severinghaus et al. (1998) (x), and probably record changes in global wetland area. Changes in the $\delta^{15}N$ values as measured by Severinghaus et al. (1998) record the temperature difference between the surface of the Greenland ice sheet and the depth at which bubbles were trapped; abrupt warmings caused the short-lived spikes in this value

data from the cores over the most recent 110,000 years, and the multiparameter analyses give an exceptionally clear view of the climate system. Briefly, the data indicate that cooling into the Younger Dryas occurred in a few prominent decade(s)-long steps, whereas warming at the end of it occurred primarily in one especially large step (Figure 1.2) of about 8°C in about 10 years and was accompanied by a doubling of snow accumulation in 3 years; most of the accumulation-rate change occurred in 1 year. (This matches well the change in wind-driven upwelling in the Cariaco Basin, offshore Venezuela, which occurred in 10 years or less [Hughen et al., 1996].)

Ice core evidence also shows that wind-blown materials were more abundant in the atmosphere over Greenland by a factor of 3 (sea-salt, submicrometer dust) to 7 (dust measuring several micrometers) in the Younger Dryas atmosphere than after the event (Alley et al., 1995b; Mayewski et al., 1997) (Figure 2.1). Taylor et al. (1997) found that most of the change in most indicators occurred in one step over about 5 years at the end of the Younger Dryas, although additional steps of similar length but much smaller magnitude preceded and followed the main step, spanning a total of about 50 years. Variability in at least some indicators was enhanced near this and other transitions in the ice cores (Taylor et al., 1993), complicating identification of when transitions occurred and emphasizing the need for improved statistical and analytical tools in dealing with abrupt climate change. Beginning immediately after the main warming in Greenland (by less than or equal to 30 years), methane rose by 50 percent over about a century; this increase included tropical and high-latitude sources (Chappellaz et al., 1997; Severinghaus et al., 1998; Brook et al., 1999).

---

near the end of the Younger Dryas and near 14.7 thousand years. Highs in sea-salt sodium indicate windy conditions from beyond Greenland, and even larger changes in calcium from continental dust indicate windy and dry or low-vegetation conditions in the Asian source regions (Mayewski et al., 1997; Biscaye et al., 1997). Calcium and sodium concentrations measured in the ice have been converted to concentrations in the air over Greenland, and are displayed by dividing by the estimated average atmospheric concentrations over Greenland in the millennium before the Little Ice Age, following Alley et al. (1997). Most of the ice-core data, and many related data sets, are available on The Greenland Summit Ice Cores CD-ROM, 1997, National Snow and Ice Data Center, University of Colorado at Boulder, and the World Data Center-A for Paleoclimatology, National Geophysical Data Center, Boulder, Colorado, www.ngdc.noaa.gov/paleo/icecore/greenland/summit/index.html. Figure is modified from Alley (2000).

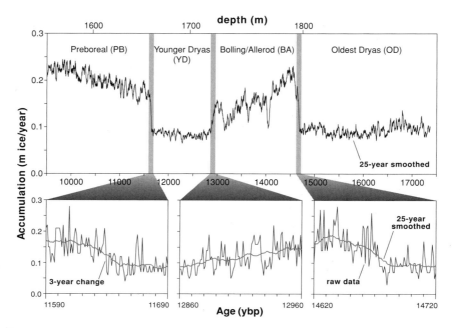

FIGURE 2.2   The accumulation rate of ice in Greenland was low during the Younger Dryas, and both the start and end of the period show as abrupt changes. Modified from Alley et al. (1993).

Ice cores from other sites, including Baffin Island, Canada (Fisher et al., 1995), Huascaran, Peru (Thompson et al., 1995), and Sajama, Bolivia (Thompson et al., 1998), show evidence of a late-glacial reversal that is probably the Younger Dryas, although the age control for these cores is not as accurate as for cores from the large ice sheets. The Byrd Station, Antarctica, ice core and possibly other southern cores (Bender et al., 1994; Blunier and Brook, 2001) indicate a broadly antiphased behavior between the high southern latitudes and much of the rest of the world, with southern warmth during the Younger Dryas interval (see Plate 2). The record from Taylor Dome, Antarctica, a near-coastal site, appears to show a slight cooling during the Younger Dryas, although details of the synchronization with other ice cores remain under discussion (Steig et al., 1998). The Southern Hemisphere records are not comparable with those from central Greenland in time resolution; further coring is planned.

The ice-core records demonstrate that much of the earth was affected simultaneously by the Younger Dryas, typically with cold, dry, windy con-

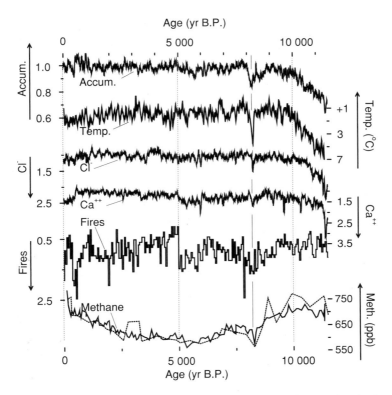

FIGURE 2.3 Climate data from the GISP2 core, central Greenland, showing changes about 8,200 years ago probably caused by outburst flooding from around the melting ice sheet in Hudson Bay (Barber et al., 1999) and affecting widespread regions of the globe. The event punctuated generally warm conditions not too different from recently, so warmth is not a guarantee of climate stability. Accumulation and temperature reflect conditions in Greenland, chloride is wind-blown sea-salt from beyond Greenland, and calcium is continental dust probably from Asia (Biscaye et al., 1997). Forest-fire smoke likely is from North America, and methane probably records global wetland area. Data are shown as approximately 50-year running means. Accumulation from Alley et al. (1993) and Spinelli (1996), chloride and calcium from O'Brien et al. (1995), and fire data shown as a 50-year histogram of frequency of fallout from fires (Taylor et al., 1996), expressed as ratios to their average values during the approximately 2,000 years just prior to the Little Ice Age. Temperature is calculated as a deviation from the average over the same 2,000 years, from oxygen-isotopic data of ice (Stuiver et al., 1995), assuming a calibration of 0.33 per mil per degree C (Cuffey et al., 1995). Methane concentrations from the GISP2 core (heavier line; Brook et al., 1996) and the GRIP core (Blunier et al., 1995) are shown in parts per billion by volume (ppb). Note that some scales increase upward and others downward, as indicated, so that all curves vary together at the major events. Modified from Alley et al. (1997).

ditions. However, those records do not provide much spatial detail, nor do they sample the whole earth. For those, one must consider a global array of data sources of various types, as described in the following subsections.

Terrestrial Pollen Evidence of the Younger Dryas

The Younger Dryas was first discovered by studying the biological records found in terrestrial sediments. These records clearly reveal the global reach of the event. Owing to dating uncertainties, including those associated with the conversion of radiocarbon measurements to calendar years, the phasing of events between different locations is not known exactly. The ice cores show that much of the world must have changed nearly simultaneously to yield the observed changes in methane, Asian dust, and Greenland conditions, but we cannot say with confidence whether all events were simultaneous or some were sequential. A summary of much of the relevant terrestrial pollen information follows, organized by region.

Europe

As the Northern Hemisphere was recovering from the last ice age about 15,000 years ago, the climate warmed dramatically and trees started to colonize the landscape. Evidence of the warming was first found in Scandinavia by geologists who noticed tree fossils in organic sediment. They named the warming interval the Allerød for the locale where it was first observed. Overlying the Allerød layer were leaves and fruits of *Dryas octopetala*, an arctic-alpine herb, in sandy or silty (minerogenic) layers above the peaty tree remains; this suggested that the climate had reverted several times to very cold conditions. Two such reversals to frigid conditions were named the Older and Younger Dryas (Jansen, 1938). Considerable evidence of this sequence in hundreds of pollen diagrams throughout Europe (Iversen, 1954; Watts, 1980) brought attention to the strongest effects of the event, which occurred in coastal Europe. During the Younger Dryas, pollen of tundra plants, such as *Artemisia* (wormwood) and *Chenopodiaceae*, abruptly replaced birch and even conifer pollen (e.g., Lowe et al., 1995; Walker, 1995; Renssen and Isarin, 1998; Birks and Ammann, 2000). In Norway, mean July temperature was about 7-9°C lower than today and about 2-4°C lower than the preceding warm Allerød interval (Birks and Ammann, 2000). It is now apparent that regional climate changes were also large in southern Europe (Lowe and Watson, 1993; Beaulieu et al., 1994). For example, mean July temperatures in northern Spain might have been as much as 8°C lower than today (Beaulieu et al., 1994).

North America

For many years, the Younger Dryas was thought to be a solely European event (Mercer, 1969; Davis et al., 1983). It was the high-resolution re-examination of pollen stratigraphy, the identification of plant macrofossils, and the new technique of accelerator mass spectrometry ^{14}C dating of these macrofossils that enabled documentation of the event in the southern New England region of the United States (Peteet et al., 1990, 1993) and in the eastern maritime provinces of Canada (Mott, 1994; Mayle et al., 1993). The climate signal in southern New England was a 3-4°C July cooling; in eastern Canada, a cooling of 6-7°C is estimated (from pollen). Midge fly fossils in lake sediments from the White Mountains of New Hampshire indicate about 5°C Younger Dryas cooling of maximum summertime lake temperatures, a somewhat smaller change than suggested for a coastal transect from Maine to New Brunswick (Cwynar and Spear, 2001). In the central Appalachians, a warm, wet interval coincident with the Younger Dryas event suggests a sharp climatic gradient that might have forced the northward movement of storm-track moisture (Kneller and Peteet, 1999). Later North American studies have identified the Younger Dryas event in other regions, such as the US Midwest (Shane and Anderson, 1993), coastal British Columbia (Mathewes, 1993) and coastal Alaska (Peteet and Mann, 1994). The documentation of the Younger Dryas event over much of North America demonstrated that it was not limited to the circum-Atlantic region (Peteet et al., 1997).

Central America and the Caribbean

Marine evidence of the Younger Dryas event is recorded as an interval of increased upwelling or decreased riverine runoff from adjacent South American land in a core from the Cariaco Basin in the Caribbean (Hughen et al., 1996, 2000a,b; Peterson et al., 2000) (Figure 2.4). Terrestrial evidence is primarily from three sites (Leyden, 1995). Evidence indicates a temperature decline of 1.5-2.5°C during deglaciation, probably correlated with the Younger Dryas, registered at high and low elevations about 13,100-12,300 years ago as far south as Costa Rica, and just before 12,000 years ago in Guatemala (Hooghiemstra et al., 1992; Leyden et al., 1994). The

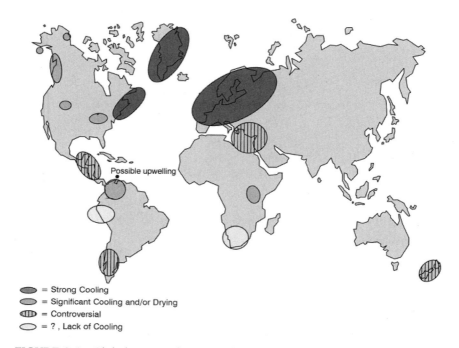

FIGURE 2.4   Global extent of terrestrial (pollen) and ice core (isotopic) evidence where the Younger Dryas cooling (11,500 – 13,000 BP) has been found. While northern hemispheric evidence is consistently strong for cooling, southern hemispheric sites contain controversial evidence and in some cases lack of evidence for a cooling during the YD interval. Possible upwelling in the Cariaco Basin during this time is also indicated, attributed to trade wind increase. Strong cooling ranges from 13-4° C; controversial means some sites show cooling and some do not (after Peteet, 1995).

decrease was not observed on the western Panamanian slope (Piperno et al., 1990; Bush et al., 1992).

South America

In Colombia, the El Abra stadial (a Younger Dryas equivalent) was a cold interval about 13,000-11,700 years ago characterized by low temperature and low precipitation (van der Hammen and Hooghiemstra, 1995). The upper forest line during the stadial was 600-800 m lower than today, and average temperatures were about 4-6°C lower than today. This evidence comes from about 14 areas, mostly at high elevations (2000-4000 m) in the Eastern and Central Cordillera and in the Sierra Nevada de Santa Marta; some data were collected from the tropical lowlands.

Late-glacial records from Ecuador do not exhibit evidence of a climatic reversal (Hansen and Sutera, 1995). Several sites in Peru give indications of a late-glacial climatic reversal although sediments from Laguna Junin indicate that the cooling occurred between 14,000-13,000 years ago, before what is normally observed for the Younger Dryas event (Hansen and Sutera, 1995). Further radiocarbon dating accompanied by high-resolution sampling is necessary. As noted above, ice cores from Peru and Bolivia show a strong late-glacial reversal (Thompson et al., 1995, 1998) that is probably correlative with the Younger Dryas, but dating is not yet unequivocal.

For several decades, southern South America has been a controversial region with respect to a possible Younger Dryas signal (Heusser, 1990; Markgraf, 1991; Denton et al., 1999). Two recent studies continue the debate from different regions of southern Chile. A study in the Lake District (Moreno et al., 2001) describes three sites at which conditions approached modern climate by about 15,000 years ago followed by cooling in two steps and then by warming around 11,200 years ago in a pattern similar to that in Europe and Greenland. The rough synchronism between Northern and Southern Hemispheres argues for a common forcing or rapid transmission of a climate signal between hemispheres. In contrast, a study farther south of four lakes shows no Younger Dryas signal (Bennett et al., 2000).

New Zealand

Late-glacial pollen evidence from New Zealand shows no substantial reversal of the trend toward warmer conditions after deglaciation (McGlone et al., 1997; Singer et al., 1998). However, a later study (Newnham and Lowe, 2000) found an interval of cooling that began about 600 years before the Younger Dryas and lasted for about a millennium; also, as noted below, one New Zealand glacier advanced near the start of the Younger Dryas interval (Denton and Hendy, 1994; cf. Denton et al., 1999).

Africa

Data from Central Africa suggest that arid conditions characterized the Younger Dryas in both highlands and lowlands (Bonnefille et al., 1995). The research focused on a high-resolution record from Burundi and compared data from 25 additional sites with limited sampling resolution and ^{14}C dating. Similarly, evidence of dry conditions during the Younger Dryas is summarized by Gasse (2000) for equatorial regions, subequatorial West Africa, and the Sahel. In South Africa, however, no strong terrestrial evidence of changes in temperature or moisture during the Younger Dryas was observed (Scott et al., 1995).

Glacial-Geological Evidence of the Younger Dryas

Glaciers are highly responsive to rapid climate change. Notable Younger Dryas advances of Norwegian and Finnish outlet glaciers and those in the Scottish mountains have been documented (Mangerud, 1991; Sissons, 1967). In the Americas, potential glacial evidence of the Younger Dryas event was observed near the Crowfoot glacier in Canada (Osborne et al., 1995; Lowell, 2000), the Titcomb Lakes moraine in the Wind River range in Wyoming (Gosse et al., 1995), and the Reschreiter glacier in Ecuador. More recent research suggests that the Younger Dryas in Peru was marked by retreating ice fronts, probably driven by a reduction in precipitation (Rodbell and Seltzer, 2000). In New Zealand, the Franz Joseph glacier began advancing early in the Younger Dryas (Denton and Hendy, 1994).

Marine Evidence of Younger Dryas Oscillation

The first evidence of Younger Dryas cooling in marine sediment cores was the observation of a return to increased abundance of the polar planktonic foraminiferal species *Neogloboquadrina pachyderma* in the North Atlantic (Ruddiman and McIntyre, 1981). This change suggested that reduction in formation of North Atlantic deep water was responsible for the Younger Dryas cooling observed on land (Oeschger et al., 1984; Broecker et al., 1985; Boyle and Keigwin, 1987). Later work documented North Atlantic ice-rafting events that correlate with rapid climate oscillations in Greenland, not only during the glacial period but also throughout the Holocene (Bond and Lotti, 1995). Deep-water corals from Orphan Knoll in the North Atlantic show large changes in intermediate-water circulation during the Younger Dryas (Smith et al., 1997). Cadmium:calcium ratios in shells from the North Atlantic subtropical gyre indicate increased nutrient concentrations during the Younger Dryas and the glacial period, and suggest millennial-scale oscillations affecting climate (Marchitto et al., 1998). Sediment color and other data from the Cariaco Basin in the Caribbean indicate enhanced nutrient upwelling and thus higher productivity caused by increased trade wind strength during the Younger Dryas (Hughen et al., 1996), or decreased riverine runoff from adjacent land masses (Peterson et al., 2000).

In the last decade, substantial paleooceanographic oscillations correlated with the Younger Dryas have been documented from as far away as the North Pacific. In the Santa Barbara Basin (Kennett and Ingram, 1995) and the Gulf of California (Keigwin and Jones, 1990), sediments that are normally anoxic became oxic during the Younger Dryas. Evidence of rapid climate variability in the northwestern Pacific over the last 95,000 years has been observed (Kotilainen and Shackleton, 1995). Even the eastern equatorial Pacific has yielded a Younger Dryas event determined from $\delta^{18}O$ and $\delta^{13}C$ records (Koutavas and Lynch-Steiglitz, 1999).

In the North Arabian Sea and Indian Ocean, high-frequency climate variability linked to events in the Northern Hemisphere has also been demonstrated (Schulz et al., 1998). Off the coast of Africa at Ocean Drilling Program Site 658, an arid period corresponding to the Younger Dryas punctuated a longer humid period (deMenocal et al., 2000a). Between 20°N and 20°S, Younger Dryas cooling is observed on the basis of alkenone paleothermometry (Bard et al., 1997). In a sediment record that links land to ocean, Maslin and Burns (2000) documented evidence of a dry Younger Dryas in the tropical Atlantic Amazon Fan. As reviewed by Boyle (2000), work including that by Boyle and Keigwin (1987) and Bond et al. (1997) showed that changes in proxies from bottom-dwelling foraminiferal shells indicate reduction in deep export of waters that sank in the North Atlantic during the Younger Dryas. Alley and Clark (1999) reviewed evidence from several marine cores that show warmth during the Younger Dyras in the southern Atlantic and Indian Oceans, opposite to most global anomalies but consistent with the warmth indicated in most Antarctic ice cores at that time (Steig et al., 1998; Bender et al., 1999; Blunier and Brook, 2001).

Overall, the available data indicate that the Younger Dryas was a strong event with a global footprint. Available data are not sufficient to identify the climate anomaly everywhere, and further understanding almost certainly will require more data. Different paleoclimatic recorders respond to different aspects of the climate system with different time resolution, so it is not surprising that the picture is not perfectly clear. Broadly, however, the Younger Dryas was a cold, dry, and windy time in much of the world although with locally wetter regions probably linked to storm-track shifts. The far southern Atlantic and many regions downwind in the southern

Indian Ocean and Antarctica were warm during the Younger Dryas. Changes probably were largest around the North Atlantic and probably included reduced export of North Atlantic deep water. Changes into and especially out of the event were very rapid.

ABRUPT CLIMATE CHANGES BEFORE
THE YOUNGER DRYAS EVENT

The 110,000-year-long ice-core records from central Greenland (Johnsen et al., 1997; Grootes and Stuiver, 1997) confirmed that the Younger Dryas was one in a long string of large, abrupt, widespread climate changes (Figure 2.5). To a first approximation, the Younger Dryas pattern of change (size, rate, extent) occurred more than 24 times during that interval; additional evidence from marine sediments indicates similar changes over longer times in earlier ice-age cycles (McManus et al., 1998).

Such climate oscillations have a characteristic form consisting of gradual cooling followed by more abrupt cooling, a cold interval, and finally an abrupt warming. Events were most commonly spaced about 1,500 years apart, although spacing of 3,000 or 4,500 years is also observed (Mayewski et al., 1997; Yiou et al., 1997; Alley et al., 2001). The name Dansgaard/Oeschger oscillation is often applied to such changes on the basis of early work by Dansgaard et al. (1984) and Oeschger et al. (1984). The terminology can be inconsistent; the warm times associated with these during the ice age originally were termed Dansgaard/Oeschger events, but evidence of cyclic behavior suggests that oscillation is more appropriate.

The sequence of Dansgaard/Oeschger oscillations is observed in various records, such as the histories of surface-water temperatures near Bermuda (which were cold when Greenland was cold) (Sachs and Lehman, 1999); oxygenation patterns of the bottom waters in the Santa Barbara basin (which were oxygenated when Greenland was cold) (Behl and Kennett, 1996); wind-blown dust supply to the Arabian Sea (which was dusty when Greenland was cold) (Schulz et al., 1998); and temperature records from the Byrd ice core, West Antarctica (which was warm when Greenland was cold) (Blunier and Brook, 2001). Methane decreased with almost all the Greenland coolings and rose with the warmings, although it changed more slowly than temperature (Chappellaz et al., 1997; Brook et al, 1999; Dällenbach et al., 2000). The colder phases of Dansgaard/Oeschger oscillations in the North Atlantic were marked by increased ice rafting of debris into colder, fresher surface water and by reduction in the strength of

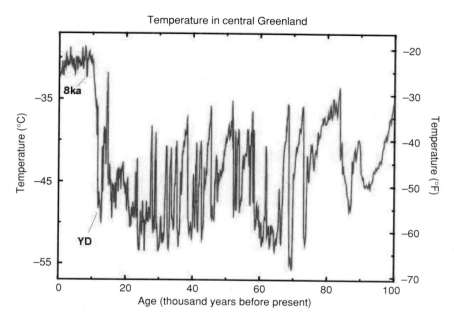

FIGURE 2.5 History of temperature in central Greenland over the last 100,000 years, as calculated by Cuffey and Clow (1997) from the data of Grootes and Stuiver (1997). The large Younger Dryas temperature oscillation (labeled YD), and the smaller temperature change of the event about 8,200 years ago (labeled 8ka) are just the most recent in a long sequence of such abrupt temperature jumps. Changes in materials from beyond Greenland trapped in the ice cores, including dust and methane, demonstrate that just as for the YD and 8ka events, the earlier events affected large areas of the earth nearly simultaneously.

North Atlantic deep water formation (e.g., Lehman and Keigwin, 1992; Oppo and Lehman, 1995; Bond et al., 1993; Bond and Lotti, 1995). The geographic pattern of climate anomalies associated with the cold phases of the Dansgaard/Oeschger oscillations is thus quite similar to that of the Younger Dryas event.

The millennial Dansgaard/Oeschger oscillations are bundled into multi-millennial Bond cycles, although with variable spacing (Bond et al., 1993). Each Dansgaard/Oeschger oscillation is slightly colder than the previous one through a few oscillations; then there is an especially long, cold interval, followed by an especially large, abrupt warming. The latter parts of the especially cold intervals are marked by the enigmatic Heinrich layers in the North Atlantic (Heinrich, 1988).

Heinrich layers are extensive deposits of coarse-grained sediment across the North Atlantic Ocean. Much of the material in these layers is sufficiently coarse that important transport by icebergs must have occurred. Each Heinrich layer is as much as 0.5 m thick near Hudson Strait, thinning to less than 1 cm on the east side of the Atlantic (Andrews and Tedesco, 1992; Grousset et al., 1993). The ice-rafted sediments are dominated by material with geochemical signatures indicating an origin in Hudson Bay, whereas sediments between and in the thin edges of Heinrich layers include more diverse sources (Gwiazda et al., 1996a,b). Sedimentation of thicker parts of Heinrich layers was much faster than that of surrounding sediments (McManus et al., 1998) and occurred in an anomalously cold and fresh surface ocean (Bond et al., 1993).

Heinrich events are correlated with greatly reduced North Atlantic deep water formation (Sarnthein et al., 1994) and climate anomalies similar to, but larger than, those of the cold phases of the non-Heinrich Dansgaard/ Oeschger oscillations (reviewed by Broecker, 1994; and Alley and Clark, 1999).

The panoply of abrupt climate change through the cooling into and warming out of the most recent global ice age and probably earlier ice ages has not been convincingly explained. However, as reviewed later, many hypotheses exist, and there is strong evidence of change in the fundamental mode of operation of parts of the coupled system of atmosphere, ocean, ice, land surface, and biosphere.

EEMIAN RAPID CLIMATE CHANGE

Temperatures similar to those of the most recent 10,000 years have been reached during previous interglacials, which have occurred approximately each 100,000 years over the last 700,000 years in response to features of earth's orbit. Each of these interglacials was slightly different from the others, at least in part because the orbital parameters do not repeat exactly. The penultimate interglacial, about 125,000 years ago, is known by several names including the Eemian, Sangamonian, and marine isotope stage 5e (with the different terminologies originating in different disciplines or geographic regions and being broadly but not identically equivalent).

As the most recent near-equivalent of the current warm period, the Eemian is of obvious interest in learning what behavior is likely during warm times (van Kolfschoten and Gibbard, 2000). The orbital parameters for the Eemian produced somewhat more incoming solar radiation than today in high northern latitudes, bringing warmer conditions, at least during summers (Montoya et al., 1998). This probably led to major retreat of the Greenland ice sheet, which likely explains high sea levels during that interval without major changes in the West Antarctic ice sheet (Cuffey and Marshall, 2000). Ice-core records from Greenland for this interval originally were interpreted as showing extremely large and rapid climate fluctuations, but flow disturbances are now known to have occurred and affected the records (Alley et al., 1995; Chappellaz et al., 1997).

Much work remains to be done on intact records from the Eemian, but it is increasingly clear from many paleoclimatic archives that although the Eemian included important paleoclimatic variability and ended abruptly, the warm period was not as variable as the periods during the slide into and climb out of the ice age that followed. In this relative stability, the Eemian had much in common with the current warm period, the Holocene.

A comprehensive survey of Eemian paleoclimatic conditions is not yet available, but a few examples of results are highlighted here. Notable variations in Eemian conditions perhaps linked to changes in oceanic circulation were documented by Fronval et al. (1998) and Bjorck et al. (2000). North Atlantic surface-water temperature fluctuations during the Eemian may have been 1-2°C, as opposed to fluctuations of 3-4°C during the cold stage that followed immediately and a deglacial warming into the Eemian of about 7°C (Oppo et al., 1997).

European pollen records are interpreted by Cheddadi et al. (1998) as indicating one rapid shift to cooler temperatures of 6 to 10°C between 4,000 and 5,000 years after the beginning of the Eemian, followed by smaller fluctuations of 2 to 4°C and 200 to 400 mm water/yr in the following few millennia. However, Boettger et al. (2000) found that the Eemian climate as recorded in isotopic data from central Germany was relatively stable, and the Eemian climate oscillations recorded in pollen records from the Iberian Margin similarly had low amplitude (Goñi et al., 1999). Cortijo et al. (2000) found that mid-latitude North Atlantic conditions during the Eemian involved no major instabilities, but that the cooling into the following glaciation occurred abruptly in less than 400 years.

Large fluctuations reconstructed for Lake Naivasha (Kenya) from sediment characteristics and diatom assemblages bear similarities to those observed during the Holocene (Trauth et al., 2001). This is at least suggestive of a general pattern of relatively more important fluctuations in low-latitude moisture availability during warm times and high-latitude temperatures during cold times.

Overall, the Eemian is neither stable and boring, nor extraordinarily variable. Most regions for which good data are available record significant and important fluctuations, some of which were abrupt, but with reduced variability compared to during the cooling into and warming out of ice ages. Attention is especially focused on drought conditions in low latitudes rather than temperature in high latitudes.

HOLOCENE RAPID CLIMATE CHANGE

The relevance of abrupt climate change of the ice age to the modern warm climate or future warmer climates is unclear. However, although glacial and deglacial rapid shifts in temperature were often larger than those of the Holocene (the last roughly 10,000 years), Holocene events were also important with respect to societally relevant climate change (Overpeck, 1996; Overpeck and Webb, 2000). For example, there were large rapid shifts in precipitation (droughts and floods) and in the size and frequency of hurricanes, typhoons, and El Niño/La Niña events. If they recurred, these kinds of changes would have large effects on society. It is not surprising that many past examples of societal collapse involved rapid climate change to some degree (Weiss and Bradley, 2001; deMenocal, 2001a).

This section summarizes some of the compelling evidence of rapid change during the Holocene. When we view the available evidence of abrupt climate change in the Holocene, it is apparent that their temporal and spatial characteristics are poorly understood. In addition, the causes of abrupt change are not well constrained. The lack of a mechanistic understanding regarding past abrupt climatic change is one of the unsettling aspects of the state of the art.

Among the most widely investigated rapid climate events of the early to middle Holocene are two that took place about 8,200 and 4,000-5,000 years ago. The former event (Figures 2.3 and 2.4) has been recognized in Greenland ice, the North Atlantic, North America, Europe, Africa, and elsewhere and has been tied to a temporary reduction in the North Atlantic thermohaline circulation generated by late-stage melting of the North American ice sheets that released a large, abrupt meltwater flood from ice-marginal lakes through Hudson Strait to the North Atlantic (Bjorck et al., 1996; Alley et al., 1997; Barber et al., 1999; Gasse, 2000; Gasse and van Campo, 1994; Kneller and Peteet, 1999, von Grafenstein et al., 1999; Yu and Eicher, 1998; cf. Stager and Mayewski, 1997). If the mechanism for this event has been identified correctly, the event was a final deglacial, or

muted Younger Dryas-like event. Changes locally might have been as large
as 10°C in the North Atlantic, with changes of about 2°C extending well
into Europe (Renssen et al., 2001). High-resolution pollen studies show
substantial and rapid vegetation response to the event in central Europe,
with early biological changes lagging climate by less than 20 years (Tinner
and Lotter, 2001). Because so many Holocene climate records are available
and the cause of the event is rather clear, it provides an opportunity for an
especially well-documented test case of model sensitivity. The event is also
important because it punctuated a time when temperatures were similar to
or even slightly above more recent levels, demonstrating that warmth is no
guarantee of climate stability.

A less well-understood hydrologic event from wet to dry conditions,
occurring roughly 5,000 years ago, also took place during a warm period.
This event is not as well documented and suffers from less than ideal temporal resolution of available records. It is most evident in African records
(Gasse and Van Campo, 1994; Gasse, 2000), the North Atlantic (Duplessy
et al., 1992; Bond et al., 1997; deMenocal et al., 2000b; Jennings et al., in
press), the Middle East (Cullen et al., 2000), and Eurasia (Enzel et al., 1999;
Morrill et al., in review). Four mechanisms have been proposed to explain
the event, all of which could have contributed. First, it might have been
associated with a cooling in the North Atlantic, perhaps related to a slow-
down in thermohaline circulation (Street-Perrott and Perrott, 1990; Gasse
and van Campo, 1994; Kutzbach and Liu, 1997; deMenocal et al., 2000b).
Second, it might be related to a subtle (and variable) ca. 1500-year oscillation in Atlantic variability (Bond et al., 1997) of poorly understood origin,
but almost certainly involving ocean processes (Alley et al., 1999), and extending beyond the North Atlantic regions; recent work (Jennings et al., in
press and Morrill et al., in review) indicated that the spatio-temporal dimensions of this variability could be complex. Third, an abrupt shift in the
El Niño-Southern Oscillation (ENSO) system might have led to a more widespread event at about the time in question (Morrill et al., in review). Fourth,
atmosphere-vegetation feedbacks triggered by subtle changes in the earth's
orbit might have triggered the event (Claussen et al., 1999) or at least amplified it (Kutzbach et al., 1996; Ganopolski et al., 1998; Braconnot et al.,
1999).

Increasing attention is also being focused on the possibility that the ENSO
system has changed its pattern of variability, perhaps rapidly. The best-docu-
mented shift in the frequency of ENSO variability occurred in 1976 (Trenberth,
1990), and it was probably one of several shifts in frequency to occur over the

last 200 years (Urban et al., 2000). Discussion continues on the statistical significance and long-term persistence of these switches and on whether they should be considered evidence of normal oscillations, of short-lived abrupt shifts, or of long-lived abrupt climate change (e.g., Rajagopalan et al., 1999; Trenberth and Hurrell, 1999a,b). Further back in the Holocene, the ENSO system might have been dramatically different from today, with much reduced variability and fewer strong events (Overpeck and Webb, 2000; Diaz and Markgraf, 2000; Cole, 2001; Sandweiss et al., 2001; Tudhope et al., 2001). Although the time at which modern ENSO variability became established is not known, there have been several model-based efforts to explain the changes, all tied to the response of the coupled atmosphere-ocean system to small orbitally induced insolation changes (Bush, 1999; Otto-Bliesner, 1999; Clement et al., 2000, 2001). The shift to more-modern ENSO variability also might have been coincident with other earth-system changes 4,000-5,000 years ago. Sandweiss et al. (2001) suggested that ENSO events were absent or substantially different from more recently between 8,800-5,800 years ago, present but reduced between 5,800-3,200 years ago, and increased to modern levels between 3,200-2,800 years ago, that would be consistent with other data that they summarize. Rodbell et al. (1999) placed the Holocene onset of El Niños at 7,000 years ago, with the beginning of modern levels reached 5,000 years ago.

Although there are other hints of important abrupt climate changes in the Holocene record, most of them have not been studied to the degree needed to place them in a coherent context (for example, examined at multiple sites). One important observation is that the landfall frequency of catastrophic hurricanes has changed rapidly during the Holocene. For example, the period about 1,000-3,500 years ago was active on the Gulf Coast compared with the last 1,000 years and changes in North Atlantic climate could be the primary cause (Liu and Fearn, 2000; Donnelly et al., 2001a,b). The period near 1,000 years ago was also possibly marked by a substantial change in hydrologic regimes in Central and North America (Hodell et al., 1995, 2001; Forman et al., 2001).

Climate variations within the last millennium are, in general, better resolved temporally and spatially than are variations earlier in the Holocene. This is due largely to the greater availability of annually dated records from historical documents, trees, corals, ice cores and sediments, but this availability is also due to greater emphasis on the last millenium by large paleoenvironmental science programs, such as PAst Global changES (PAGES) of the International Geosphere-Biosphere Programme (IGBP). Perhaps the most studied rapid temperature shift of the Holocene is the change that began in the latter half of the nineteenth century and ended the so-called Little Ice Age. The shift and later state of substantial global warming were unprecedented in the context of the last 500 years and might be due to a combination of natural (such as solar and volcanic) and human-induced (such as trace-gas) forcing (Overpeck et al., 1997; Jones et al., 1998; Mann et al., 1998, 1999, 2000; Huang et al., 2000; Crowley, 2000; Briffa et al., 2001; Intergovernmental Panel on Climate Change, 2001a).

In contrast with the abrupt late nineteenth to early twentieth century warming, timing of the onset of the Little Ice Age is difficult to establish in that the change manifests itself as a period of slow Northern Hemisphere cooling beginning at or before ca 1000 (Mann et al., 1999; Crowley, 2000; Crowley and Lowery, 2000; Briffa et al., 2001) with several sustained cooler intervals thereafter (for example, the seventeenth century and early nineteenth century).

There are insufficient paleoclimate records to allow complete reconstruction of the last 1,000 years of change in the Southern Hemisphere, and uncertainty remains on the amplitude of Northern Hemisphere change in this interval (e.g., Briffa et al., 2001; Huang et al., 2000). There is still debate as to whether the "Medieval Warm Period" was more than a Northern Hemisphere warm event (Mann et al., 1999; Crowley, 2000; Crowley and Lowery, 2000; Briffa et al., 2001; Broecker, 2001). Moreover, evidence is scarce outside the North Atlantic-European sector (Jennings and Weiner, 1996; Keigwin, 1996; Broecker, 2001) for medieval temperatures that were close to mean twentieth century levels. Additional annually resolved records for the last 2,000 years are needed to answer such fundamental questions.

Holocene Droughts

The existing temperature records, as described above, make it clear that natural variability alone can generate regional to hemispheric temperature anomalies that are sufficient to affect many aspects of human activity. However, the record of hydrologic change over the last 2,000 years suggests even larger effects: there is ample evidence that decadal, even century-scale, drought can occur with little or no warning.

A synthesis of US drought variability over the last 2,000 years (Woodhouse and Overpeck, 1998) used records from a diverse array of proxy sources (cf. Cronin et al., 2000; Stahle et al., 1998). From this synthesis, it was concluded that multi-year droughts similar to the 1930s Dust

Bowl or the severe 1950s southwest drought have occurred an average of once or twice per century over the last 2,000 years. Furthermore, decadal "megadroughts" have also occurred often, but at less frequent intervals. The last of these occurred in the sixteenth century, spanned much of northern Mexico to Canada, and lasted over 20 years in some regions (Woodhouse and Overpeck, 1998; Stahle et al., 2000). An earlier event in the thirteenth century also persisted for decades in some locations and involved the long-term drying of lakes in the Sierra Nevada of California (Stine, 1994) and the activation of desert dunes in parts of the High Plains (Muhs and Holliday, 1995, 2001). There is evidence of even longer droughts further back than the last millennium (Stine, 1994; Laird et al., 1996; Fritz et al., 2000), including an unprecedented multidecadal drought that has been implicated in the collapse of the Classic Mayan civilization (Hodell et al., 1995, 2001), several droughts that led to the remobilization of eolian landforms on the High Plains (Forman et al., 2001), and linkage between droughts in tropical and temperate zones (Lamb et al., 1995). An important conclusion from paleodrought research is that drought regimes can shift rapidly and without warning. A prominent example is the shift, at about 1200 BP, from a regime characterized by frequent long droughts on the High Plains to the current regime of less-frequent and shorter droughts (Laird et al., 1996; Woodhouse and Overpeck, 1998).

Despite growing knowledge of the paleodrought record, causal mechanisms of changes are poorly understood (Woodhouse and Overpeck, 1998). Persistent oceanic temperature anomalies, perhaps related to ENSO or the North Atlantic Oscillation (NAO) as described below, have been proposed as one potential forcing mechanism (Forman et al., 1995; Black et al., 1999; Cole and Cook, 1998; Cole et al., submitted), but cause and effect have yet to be proved in the case of any decadal or longer paleodrought in North America. There is also good evidence of late Holocene multidecadal droughts outside North America (e.g., Stine, 1994; Verschuren et al., 2000; Nicholson, 2001); their causes are equally enigmatic. Thus, although we know that droughts unprecedented in the last 150 years have occurred in the last 2,000 years and so could occur in the future, we do not have the scientific understanding to predict them or recognize their onset.

Holocene Floods

Just as the twentieth century instrumental record is too short to understand the full range of drought, it is too short to understand how the frequency of large floods has changed (Baker, 2000). Data on past hydrological conditions from the upper Mississippi River (Knox, 2000) and from sediments in the Gulf of Mexico (Brown et al., 1999) record large, abrupt shifts in flood regimes in the Holocene, which may have been linked to major jumps in the location of the lower Mississippi (delta-lobe switching). In the western United States, there is growing evidence that flood regimes distinctly different from today, and also episodic in time, were the norm rather than the exception. The frequency of large floods in the Lower Colorado River Basin, for example, appears to have varied widely over the last 5,000 years (Ely et al., 1993; Enzel et al., 1996), with increased frequency from about 5,000-4,000 years ago, then lower frequency until about 2,000 years ago, and some abrupt shifts up, down, and back up thereafter (Ely, 1997). Those flood-frequency fluctuations and substantial fluctuations elsewhere around the world (e.g., Gregory et al., 1995; Baker, 1998; Benito et al., 1998) appear to be linked to climate shifts but in poorly understood ways. Clearly, a predictive understanding of megadroughts and large floods must await further research.

This observation about droughts and floods applies at some level to all the abrupt climate changes recorded in proxy records. The data are clear. Ice-age events were especially large and widespread and involved changes in temperature, precipitation, windiness, and so on. Holocene events were more muted in polar regions, might have been more regionalized, and usually involved water availability, but often with important temperature changes as well. Multi-characteristic global-anomaly maps are not available for any of the abrupt changes, and additional records and proxy techniques will be required to provide such anomaly maps. Coverage gaps appear especially large in the oceans and southern latitudes, although broad gaps also exist elsewhere.

RAPID CLIMATE CHANGES IN THE INSTRUMENTAL PERIOD

Instrumental records from scientific monitoring programs offer the possibility of capturing directly the relevant data on abrupt climate change with greater accuracy and spatial coverage than are possible from the necessarily limited proxy records. The relatively short period of instrumental records means that they have missed most of the abrupt changes discussed above, although some droughts and the warming from the Little Ice Age have been captured rather well. Instrumental records will become more valuable as their length increases, which argues for maintenance of key observational data sets. Instrumental records also are critical in characterizing patterns of climate variability that might have contributed to paleoclimatic abrupt change, and might contribute to abrupt climate change in the future. It is important to the understanding of abrupt climate change that these patterns or "modes" of circulation and its variability be understood, particularly on the time-scale of decades to centuries. The abrupt changes surveyed here are smaller in strength than the extreme events of the paleoclimate record, yet they are nonetheless significant as human populations press the capacity of the environment, locally and globally.

Atmospheric instrumental data include surface values and vertical profiles of numerous physical variables, including temperature, pressure, radiation, and winds. Surface observations, satellite radiometric observations, and the global network of regularly launched radiosonde profilers are assimilated into computer models of the atmosphere to analyze weather and climate. They capture both the conditions that cause atmospheric circulation and the resulting atmospheric motions. Much of our current understanding of climate comes from the relatively accurately observed period since 1950. More subtle are the measurements of trace chemicals, which both affect the physical state of the atmosphere, and can be used to infer its motions. The longest atmospheric time series, dating back several hundred years, are surface temperature and pressure.

The ocean, like the atmosphere, is a thin fluid envelope covering much of the earth. Satellites are now collecting global observations of the temperature, elevation and roughness of the sea surface, which tell us the surface currents and winds fairly accurately. Crucial climate variables, such as sea-ice cover and movement (and to a lesser accuracy, ice thickness), have been measured by satellites beginning in the 1970s. Yet, oceanic data are still more restricted in coverage and duration than atmospheric data, for it is still difficult to penetrate the depths of the ocean with instruments in sufficient numbers.

In addition to the purely instrumental problem, ocean currents and eddies are smaller in size than major atmospheric wind fields, making the mapping of ocean circulation more difficult (weather patterns are well matched in size to the spacing of major cities, which historically made their discovery possible, using simple barometers). Another contrasting property is the time for fluid to adjust fully to a change in external forcing: in the atmosphere this time is a month or two, while in the ocean it is measured in millennia. The ocean dominates the global storage of heat, carbon, and water of the climate system while the atmosphere dominates the rapid response of the climate system and more directly impacts human activity.

The ocean's direct impact on the atmosphere is primarily through sea-surface temperature and ice-cover. Thus, it is fortunate that temperature records are among the longest oceanic time-series and have the best spatial coverage. Data sets include sea-surface temperatures from ocean vessels, long coastal sea-level and temperature records, and shorter or more scattered time-series of temperature and salinity from surface to sea-floor. Increasingly long time-series of directly measured ocean currents are becoming available, particularly in the tropics. The TAO (Tropical Atmosphere-Ocean) array in the Pacific, sometimes called the world's largest scientific instrument, measures equatorial temperatures, winds and currents around one-quarter of the earth's circumference (e.g., McPhaden et al., 1998). The array has given us detailed portraits of El Niño-Southern Oscillation (ENSO) cycles and equatorial general circulation.

Over longer times other aspects of oceanic circulation, chemistry and biology become important to climate. For example, the heat storage available to the atmosphere is strongly dependent on circulation and salinity stratification of the upper ocean. The depths of the ocean become involved as the thermohaline circulation (THC) and wind-driven circulation interact to reset surface conditions. There are "overturning circulations" at many scales, from the global THC (see Plate 4) to the shallow, near-surface cells of overturning lying parallel to the equator. Direct measurements of the circulation of the deep ocean are still sparse, and indirect means are often used to infer the circulation. Water density (from measured temperature and salinity) can be combined with dynamical constraints and atmospheric observations of air-sea interaction to estimate global ocean circulation (e.g., Ganachaud and Wunsch, 2000; Reid, 1994, 1998, 2001). The results are consistent with the limited direct measurements of currents, and also with the patterns of observed chemical tracers in the ocean. The tracers include natural dissolved gases and nutrients, dynamical quantities such as potential vorticity and potential density, and chemical inputs from human activity. Transient chemical tracers, injected into the atmosphere and subsequently absorbed by the ocean, provide particularly useful images of the ocean circulation. Bomb radiocarbon, tritium and chlorofluorocarbons (CFCs), for example, allow verification and quantitative assessment of the pathways of high-latitude sinking, equatorward flow in boundary currents, and interaction with the slower flow of mid-ocean regions (Broecker and Peng, 1982; Doney and Jenkins, 1994; Smethie and Fine, 2001).

The location, strength and depth penetration of the major sinking regions of the ocean at high latitude (see Plate 4) are known to have changed during glacial cycles, emphasizing the importance of sea-ice cover in insulating the ocean from the atmosphere, preventing deep convection and physical sinking from occurring (e.g., Sarnthein et al., 1994). The contrasting effect of freezing sea-water is that salty brine is rejected from the ice, yielding a small but very dense volume of water that can contribute to sinking events. During the twentieth century lesser yet still significant shifts of the deep circulation (e.g., Molinari et al., 1998) have been verified by tracers and direct current measurements.

Abrupt changes in climate can occur with spatial patterns that in some way reflect the natural dynamics of atmosphere and ocean. These "modes" of circulation are seen in the seasonal, interannual and decadal variability of the system, and have great potential as an aid to understanding just how abrupt changes can occur. At work in establishing the modes are "teleconnections" both vertically, and across the globe. Various waves, particularly Rossby (or "planetary") waves and Kelvin waves, and unstable waves on the time-averaged circulation, are involved, as is the direct transport of climate anomalies by the circulation.

Natural variability of climate is now occurring in the context of global warming, so the discussion of abrupt climate change during the period of instrumental records must acknowledge the presence of anthropogenic and natural change, and the possibility of strong interaction between them.

PATTERNS OF CLIMATE VARIABILITY

Instrumental records show that the climate is characterized by patterns or modes of variability, such as the polar annular modes and ENSO of the equatorial Pacific, as described below. The spatial patterns can provide regional intensification of climate change in quite small geographic areas. The strong couplings and feedbacks among at least the atmosphere, oceans and sea ice, and probably other elements of the climate system, allow a pattern to persist for periods of years to many decades. The different regional modes also interact with one another. For instance, Amazonian rainfall responds to a mode of tropical Atlantic variability, which itself might be responding to ENSO or the Arctic Oscillation.

The behavior of highly idealized models of the climate system suggests that climate change can be manifested as a shift in the fraction of the time that climate resides in the contrasting phases (for example, warm/cold or strong-wind/weak-wind) of such oscillations (Palmer, 1993). However, the scientific community is divided on the issue of whether analogous "regime-like" behavior exists in instrumental records related to the real climate system. Hansen and Sutera (1995), Corti et al. (1999), and Monahan et al. (2000) found evidence of such mode-shift behavior. However observational evidence has been questioned (e.g., Nitsche et al., 1994; Berner and Branstator, 2001). Also, questions remain about whether such behavior should be characteristic of an entity with as many degrees of freedom as the climate system (Dymnikov and Gritsoun, 2001).

The possibility that mode shifts participated in or provide clues to the large, abrupt climate changes of preinstrumental times suggests common mechanisms or even common causes. Thus, the study of abrupt climate change should involve consideration of the preferred modes of the climate system.

Annular Modes

The annular modes—the Arctic Oscillation (AO) and the Antarctic Oscillation (AAO)—primarily affect polar to middle-latitude regions in both the North and South and are the dominant modes of climate variability in these areas, especially in the winter. The AO and AAO represent a transfer of atmospheric mass between subtropical high-pressure regions and polar lows. A strongly positive state of an annular mode is associated with intensified highs and lows driving strong atmospheric circulation. The negative state has much less difference between high- and low-pressure regions and thus is related to weaker atmospheric circulation.

The southern annular mode is moderately symmetric about the pole, but owing to the complex geometry of northern continents, the AO is especially strong over the North Atlantic and less evident in other regions. Thus, the mode was originally described as the North Atlantic Oscillation (NAO), and an NAO index was based on the difference in atmospheric pressure between Portugal and Iceland (Hurrell, 1995). When the winter pressure difference is large, frequent strong storms take a northeasterly track across the North Atlantic, producing warm and wet weather in northern Europe, cold and dry conditions in northern Canada, and mild and wet conditions along the US East Coast. In contrast, a small pressure difference produces fewer, weaker storms, taking an easterly track to produce a moist Mediterranean, cold northern Europe, and a snowy US East Coast in response to frequent cold-air outbursts.

ENSO and ENSO-Related Variability

A weakening of the trade winds in the equatorial Pacific and attendant warming of the sea surface (or lack of cooling by upwelled cold water) is known as an El Niño event. Such events alternate with an opposing state, popularly referred to as "La Niña," with strong trade winds and upwelling of cold waters off Peru and along the Equator. The few-year oscillation between those different states is the El Niño/Southern Oscillation. The coupled oscillation of the tropical ocean and atmosphere is important in global climate, with impacts that extend far beyond the tropical Pacific to the tropical Atlantic and Indian Oceans, to the Southern Ocean, and to middle to high latitudes in the Northern Hemisphere. There are speculations that greenhouse warming is sufficient to put the world into a warmer, near-perpetual El Niño state (e.g., Timmerman et al., 1999; Federov and Philander, 2000), but there is no strong consensus.

ENSO might be linked to another of the leading patterns of variability, the so-called Pacific North American (PNA) pattern, which exerts a strong influence on distribution of rainfall and surface temperature over western North America. Like the AO, the PNA pattern fluctuates randomly from one month to the next, but also exhibits what appear to be systematic variations on a much longer time scale. Since 1976-1977, the positive polarity of the PNA pattern—marked by a tendency toward relatively mild winters over Alaska and western Canada, below-normal rainfall and stream flows over the Pacific Northwest, and above-normal rainfall in the southwestern United States—has been prevalent, whereas during the preceding 30-year period the opposite conditions prevailed.

The abrupt shift toward the positive polarity of the PNA pattern in 1976-1977 was coincident with and believed to be caused by a widespread pattern of changes throughout the Pacific Ocean. Sea-surface temperatures along the equatorial belt and along the coast of the Americas became warmer, while farther to the west at temperate latitudes the sea surface became cooler (Nitta and Yamada, 1989; Trenberth, 1990; Graham, 1994). An array of changes in the marine ecosystem occurred around the same time (Ebbesmeier et al., 1991). For example, salmon recruitment underwent a major readjustment toward more abundant harvests along the Alaskan coast accompanied by deteriorating conditions in southern British Columbia and the US Pacific Northwest (Francis and Hare, 1994). Another basin-wide "regime shift" that was analogous in many respects to the one that occurred in 1976-1977, but in the opposite sense, was observed during the 1940s (Zhang et al., 1997; Minobe and Mantua, 1999), and there are indications of prior shifts as well (Minobe, 1997). The suite of atmospheric and oceanic changes that have been linked to these basin-wide regime shifts is collectively referred to as the Pacific Decadal Oscillation (PDO) (Mantua et al., 1999).

The sea surface temperature (SST) patterns associated with the PDO and ENSO are similar, the main distinction being that the extratropical features are somewhat more prominent in the PDO pattern. As in the few-year variations associated with the swings between El Niño (warm) and La Niña (cold) conditions in the equatorial Pacific, warm and wet decades in the equatorial zone tend to be marked by extratropical circulation patterns that favor an unusually active storm track in the mid-Pacific that splits toward its eastern end. An unusually large fraction of disturbances moves northeastward, bringing mild, wet weather to the Alaska panhandle; many of the remainder track southeastward, bringing heavy rains to southern California and the US desert Southwest. The mountain ranges of British Columbia and the US Pacific Northwest, which lie directly downstream from the split in the storm track, tend to receive less than the normal amount of winter snowfall, and this reduces water supplies for the following summer season. The dynamic mechanisms responsible for the long-range "teleconnections" between the equatorial Pacific and the extratropics are better understood than the processes that control the evolution of this phenomenon on the decadal time scale. Hence, regime shifts such as the one that occurred in 1976-1977 are difficult to diagnose in real time, let alone to predict.

There are several different schools of thought as to the nature of the interdecadal PDO variability, which has shown both the abruptness and persistence to qualify under our definition of abrupt climate change. The default hypothesis is that the PDO is merely a reflection of stochastic variability originating in the atmosphere but amplified by positive feedbacks associated with coupling between the atmosphere and ocean (Bretherton and Battisti, 2000). If this interpretation is valid, it follows that this ENSO-like variability is inherently unpredictable (i.e., that it becomes clearly evident only with the benefit of hindsight). Hopes that the phenomenon is deterministic, and therefore predictable, are based on the notion that ocean dynamics play an active role in PDO evolution, to the extent of setting the time scale for the major swings back and forth between the positive and negative polarity of the PDO pattern. One oceanic process that could conceivably set the time scale is the recirculation time for water parcels in the clockwise North Pacific and counterclockwise South Pacific subtropical gyres. A second subtropical gyre time scale is set by the time it takes for oceanic planetary waves to propagate to the western boundary currents, which then feed back on the atmospheric circulation. A third is the time required for water parcels subducted in the extratropical North and South Pacific at latitudes around 35°N and 25°S to reach the equatorial thermocline. Mechanisms that depend on those processes have been demonstrated to be capable of producing ENSO-like interdecadal variability in coupled atmosphere-ocean models (Latif and Barnett, 1996). Further data and model results are needed to learn the extent to which the time scales of the variability can change and whether the climate can "lock into" one or another phase of the major oscillations. Mean ice-age conditions in the tropical Pacific appear to have been more La Niña-like than during the Holocene; perhaps this suggests a linkage. Species and shell chemistry and isotopic ratios of planktonic foraminfera (Lee et al., 2001) and chemistry and isotopic ratios of corals (Tudhope et al., 2001) give evidence for equatorial Pacific sea-surface temperatures back at least 130,000 years. Cooler mean SST during the glaciations (~3°C cooler than modern at the last glacial maximum in the Lee et al. study; also, Patrick and Thunell, 1997; Pisias and Mix, 1997; also see Alley and Clark, 1999) and continued, yet weaker ENSO cycles are evident. Stronger glacial easterly equatorial winds are inferred (Lyle, 1988).

Tropical Variability in the Atlantic and Indian Oceans

Tropical variability arising from feedbacks within the Atlantic and Indian equatorial regions also contributes to regional climate modes, although of smaller global impact than ENSO, probably because of the vast width of the Pacific relative to the Atlantic or Indian. Tropical Atlantic variability correlates strongly with forcing from ENSO and the AO. The tropical Atlantic also has a mode that is symmetric about the equator with mechanisms similar to those in ENSO and might contribute to regional predictability (Amazonian and west African/Sahelian rainfall). Off-equatorial modes of tropical Atlantic variability are associated with the strength and location of the northern and southern Intertropical Convergence Zones (ITCZs); work in recent years has revealed that Northern and Southern Hemisphere SST variability are not tightly linked. Tropical Atlantic variability has a major impact on rainfall in northern Africa and northern South America and an impact on hurricane frequency and patterns in the North Atlantic.

In the Indian tropical region, the seasonal monsoon driven by ocean-land temperature contrasts has a major impact on human life. The monsoon is perhaps the classic example of ocean-atmosphere-land interactions. During boreal summer, a northward shift of the ITCZ to the Indian subcontinent creates a major precipitation and heat source in this region. Interannual variability in the Indian monsoon correlates closely with tropical Indian Ocean SST. Indian Ocean SST is affected by ENSO and by an intrinsic Indian Ocean east-west mode of variability similar in mechanism but uncorrelated with the Pacific's ENSO.

Extended Summer Drought

The Northern Hemisphere's annular mode and the decade-to-decade ENSO-like variability discussed in the previous sections both affect Northern Hemisphere climate mainly during the winter season, and they involve the atmosphere's own preferred modes of month-to-month variability. In contrast, drought and desertification, when they occur in extratropical latitudes, are primarily summer phenomena whose geographic distribution and evolution are determined as much by land-surface processes as by atmospheric dynamics. Dynamical modes may still be involved, however, as the summer pattern of great anticyclones over the oceans responds to the heating of the continents. Kelvin and Rossby waves are active in determining the shape, extent, and flow of moisture in this pattern (Rodwell and Hoskins, 2001), and in turn these waves are involved in dynamical modes as noted above.

An extended drought popularly known as the Dust Bowl affected large areas of the United States through most of the decade of the 1930s. Over parts of the Great Plains and Midwest, the 1931-1939 summers were on the average substantially warmer than the long-term climatological mean for the season, with daily maxima often in excess of 40°C, and precipitation was deficient (Borchert, 1950; Skaggs, 1975; Karl and Quayle, 1981; Diaz, 1983; Chang and Wallace, 1987; Chang and Smith, 2001). Much of the topsoil was irreversibly lost—blown away in dust storms that darkened skies as far downstream as the eastern seaboard. Numerous farms were abandoned, and agricultural productivity dropped sharply. Many who lived through the Dust Bowl must have wondered whether climatic conditions would ever be suitable for farming again. Yet toward the end of the decade, the rains returned, and the region has never since been plagued by such an extended drought.

What initiated the Dust Bowl in the early 1930s and what caused the rains to return nearly a decade later are still open questions. The prevailing view is that drought is an inherently stochastic phenomenon, initiated and terminated by random fluctuations in atmospheric circulation patterns, and sustained over long periods of time by positive feedback from the terrestrial biosphere (Namias, 1960; Rind, 1982; Shukla and Mintz, 1982; Karl, 1983; Sud and Molod, 1988; Bravar and Kavvas, 1991; Xue and Shukla, 1993; Dirmeyer, 1994; Lare and Nicholson, 1994). A few weeks of abnormally hot, dry weather are sufficient to desiccate the upper layers of the soil, reducing the water available for plants to absorb through their root systems. The plants respond by reducing the rate of evapotranspiration through leaves during the daylight hours (Dirmeyer, 1994; Radersma and de Reider, 1996; Xue et al., 1996). Reduced evapotranspiration inhibits the ability of the plants to keep themselves and the earth's surface beneath them cool during midday, when the incoming solar radiation is strongest (Somayao et al., 1980; Gardner et al., 1981). This favors higher afternoon temperatures and also reduces the humidity within the lower 1-2 km of the atmosphere (Walsh et al., 1985; Karl, 1986; Georgakakos et al., 1995; Huang et al., 1996; Dai et al., 1999). Because this boundary-layer air is the source of roughly half the moisture that condenses in summer rainstorms over the central United States, lower humidity favors reduced precipitation (Brubaker et al., 1993; Eltahir and Bras, 1996; Koster and Suarez, 1996; Findell and Eltahir, 1999; Trenberth, 1999). Higher daily maximum temperatures, lower humidity, and reduced precipitation all increase the stress on plants. If the stress is sufficiently severe and long, the physiological changes in plants become irreversible. Once the threshold is crossed, the earliest hope for the restoration of normal vegetation is the next spring growing season, which can be 6 or even 9 months away. Throughout the remainder of the summer and early autumn, the parched land surface continues to exert a feedback on the atmosphere that perpetuates the abnormally hot, dry weather conditions (Yeh et al., 1984; Huang and Van den Dool, 1993; Yang et al., 1994; Huang et al., 1996; Fennesy and Shukla, 1999).

The wilting of the plants also affects hydrological conditions in the ground. In the absence of healthy root systems, water runs off more rapidly after rainstorms, leaving behind less to nurture the plants. Once the water table drops substantially, an extended period of near- or above-normal precipitation is required to restore groundwater (Palmer, 1965; Entekhabi et al., 1992; Bravar and Kavvas, 1991; Stamm et al., 1994). The remarkable year-to-year persistence of the 1930s drought attests to the memory of the vegetation and the ground. Once established, an arid climate regime, such as the one that prevailed during the Dust Bowl, appears to be capable of perpetuating itself until a well-timed series of rainstorms enables the vegetation to regain a foothold (Dirmeyer and Shukla, 1996; Wang and Eltahir, 2000a,b; Clark et al., 2001).

The onset and termination of the 1930s Dust Bowl are examples of abrupt regime shifts from a climate conducive to agriculture to a climate more characteristic of a desert region and back again. During the time covered by instrumental records, such shifts have occurred rather infrequently in the United States but more regularly in semiarid agricultural regions, such as the Sahel, northeast Brazil, and the Middle East (Nicholson et al., 1998; Street-Perrott et al., 2000). If such dry regimes are sufficiently frequent or long, the cumulative loss of topsoil due to wind erosion makes it increasingly difficult for vegetation to thrive, and difficult-to-reverse "desertification" occurs (United Nations, 1980). Thus far, the United States has experienced relatively little true desertification, but other regions of the globe have not been as fortunate. For example, it is well documented that the Sahara expanded northward and engulfed formerly productive agricultural regions of North Africa during the last few centuries of the Roman Empire (Reale and Dirmeyer, 2000); this transition might well have involved a series of prolonged drought episodes analogous to the Dust Bowl.

Agricultural practices influence the retention of topsoil. Poor cultivation practices and overgrazing have been blamed for the desertification that has plagued North Africa, the Sahel, and other semi-arid regions (Otterman 1981; Wendler and Eaton, 1983; Balling, 1988; Bryant et al., 1990; Ben-Gai et al., 1998; Nicholson et al., 1998; Pickup, 1998), and the planting of hedgerows designed to impede the flow of wind-blown dust has been credited with sparing much of the US Great Plains from suffering a similar fate. Whether adherence to environmentally sound agricultural practices will be sufficient to prevent further desertification is less clear.

Global warming could render such regions as the western and central United States more vulnerable to extended drought episodes by increasing temperatures during the growing season, and thereby increasing the rate of evapotranspiration. There is no conclusive evidence of such behavior in response to the rapid warming of the last two decades, but simulations with climate models indicate that more pronounced warming like that predicted to occur by the end of the twenty-first century could serve to increase the frequency of drought episodes and the risk of irreversible desertification (Rind et al., 1989; Henderson-Sellers et al., 1995; Bounoua et al., 1999).

TRENDS RECORDED INSTRUMENTALLY

Instrumental records are becoming long enough to show trends in climate, although the distinction between trends and longer cycles is never simple. The Little Ice Age set the stage for the global warming of the last century, which probably had both natural and anthropogenic causes (Intergovernmental Panel on Climate Change, 2001b). A brief review of some of the main trends is provided here, showing that these trends often exhibit regional abruptness and are linked to modes of climate variability, again documenting the importance of instrumental records to the study of abrupt climate change.

Most long instrumental records were initiated in the nineteenth and twentieth centuries. The central England temperature time series is one of the longest continuous instrumental climate records (Figure 2.6), and it cap-

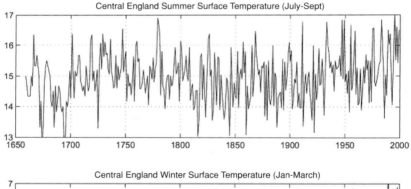

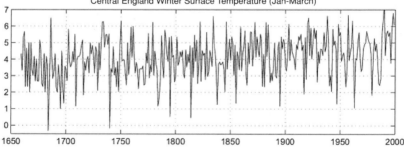

FIGURE 2.6   The Central England temperature time-series is one of the longest continuous instrumental records of climate change, with data from 1659 to 2000. Top figure shows summer and bottom figure shows winter. Temperature scale is in degrees Celcius. (Data courtesy of P. Jones, University of East Anglia.)

tured not only the warming of the last century, but also much of the preceding Little Ice Age (Jones and Bradley, 1992). Long cool periods, such as the summers throughout the nineteenth century, were punctuated by extreme events. The cold phases were known to have had important effects on human activity, as in the abrupt European cooling from the unusually warm 1730s to the cold 1740s. Famine occurred across western Europe, especially in Ireland and France, where farmers who depended on wheat and potatoes were slow to adapt. In Ireland, this is known as the "forgotten famine"; as many people died as in the famed "potato famine" of the 1840s.

In much of the northern United States, 1816 was the "year without a summer" (Stommel and Stommel, 1983); it was linked to the atmospheric veil from the volcanic eruption of Tambora in Indonesia. Snow fell in July in northern New England, farming was disrupted, and crops failed. The event might have spurred the westward migration of farmers from the thin soils of the hill farms of New Hampshire. Without suggesting a causal link, La Niña occurred in late 1815; by late 1816, a strong El Niño developed. There was severe drought in Brazil in 1816-1817. Single volcanic eruptions often have a widespread cooling effect (e.g., Grootes and Stuiver, 1997; Zielinski et al., 1997). Such effects ordinarily would be considered "noise" in a climate state; however the compounding of such noise with longer climate trends, such as the Little Ice Age, can increase the impacts, as observed. Volcanic forcing also represents an experiment in which the response of global climate modes can be observed in great detail, helping understand how modes might respond to other forcings.

AO/NAO Trends and Variability

The winter atmospheric circulation in the Northern Hemisphere has undergone some remarkable changes during the last several decades. Sea-level pressures over the Arctic have fallen by about 6 hPa (Walsh et al., 1996), and the subpolar westerly winds have strengthened, particularly over the Atlantic sector. Related circulation changes have favored mild winters over most of Russia, China and Japan, and drought over southern Europe and parts of the Middle East (Hurrell, 1995; Thompson et al., 2000). Warming during the period has been concentrated in central and northern Asia and northwestern North America. Retreating glaciers, warming permafrost, and decreasing sea-ice cover have been observed in Alaska, where temperatures increased abruptly in the late 1970s, distinct from the post-1980 acceleration of globally averaged temperature rise. The patterns in these trends bear a strong resemblance to the Northern Hemisphere's leading pattern of variability, the Arctic Oscillation/North Atlantic Oscillation (AO/NAO) (Thompson and Wallace, 1998; Wallace, 2000). That is, the observed trends in the wintertime temperatures and precipitation appear to be largely regional expressions of the positive trend of the AO/NAO. There are indications of analogous trends in the Southern Hemisphere, in association with its annular mode (Thompson et al., 2000).

Several ideas have been put forth concerning the cause of those trends. Shindell et al. (1999, 2001) have proposed that they might be occurring in response to increasing concentrations of greenhouse gases, which warm the lower atmosphere but cool the stratosphere. Indeed, similar trends have been simulated in an atmospheric model driven by increasing concentrations of greenhouse gases. It has also been suggested that stratospheric ozone depletion (Volodin and Galin, 1999) and trends in sea-surface temperatures (Rodwell et al., 1999; Hoerling et al., 2001) might be capable of inducing similar trends in the AO/NAO. Because favored patterns of variability tend to be highly sensitive to external forcing, it is indeed quite plausible that they can be changed in a number of ways. Fyfe et al. (1999) give a contrasting view, that the AO/NAO did not appear to be altered substantially by greenhouse forcing, yet its Southern Hemisphere counterpart, the annular "AAO" mode, was indeed strongly amplified.

Whatever its cause, the observed trend in the AO/NAO is influencing other components of the earth system. The trend has altered the distribution of deep convection over the North Atlantic (Lilly et al., 1999, Dickson et al., 1996), caused glaciers in Norway to advance (Hurrell, 1995), and contributed to the thinning of the stratospheric ozone layer over middle and high latitudes of the Northern Hemisphere during winter and early spring (Thompson et al., 2000). Interactions with the oceanic thermohaline circulation (THC) have the potential to exert positive feedback on the AO/NAO itself, increasing its variability on decadal and longer scales. Owing to such feedback, it is conceivable that shifts in the AO/NAO contributed to some of the large climatic swings in the paleoclimate record.

If trends observed in the AO/NAO during the last 30 years were to continue into the twenty-first century, they could have important societal impacts. For example, the recovery of the stratospheric ozone layer from human-induced depletion during the twentieth century could be delayed by several decades (Shindell et al., 1999).

Abrupt Warming of the North Atlantic, 1920 to 1930

Global-average surface temperature records show two principal periods of warming during the twentieth century. The warming following 1970 has occurred widely, yet with regions of concentration in northern Asia and northwestern North America. Anthropogenic forcing is widely suspected to be a contributing cause (Intergovernmental Panel on Climate Change, 2001b). The warming earlier in the century is less likely to be of anthropogenic origin, although this is an unsettled issue; solar and volcanic forcing may have played a role (Delworth and Knutson, 2000; Stott et al., 2000; Hegerl et al., 2000).

Global temperature has evolved very differently, in space, during the two warmings (Figure 2.7). The earlier twentieth-century episode was concentrated in the far north. The warming seems to have appeared first in the Barents Sea; for example, in the records from Turuhansk, a city on the Yenisey River in northern Siberia. Sparse temperature records are supplemented by observations of a strong increase in ocean salinity and decrease in sea ice cover in the Nordic Seas (Kelley et al., 1982, 1987). Good records from Danish stations on the west Greenland coast show the strongest and most abrupt arrival of warming at Upernavik, where a 4°C rise in surface

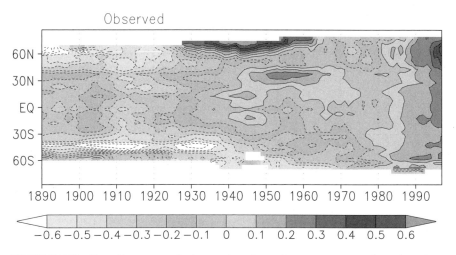

FIGURE 2.7 Zonally averaged observed surface air temperature, plotted against latitude and time. Land stations only (Delworth and Knutson, 2000). Reading figure from left to right, one sees the rapid, concentrated arrival of warm temperatures in the 1920s, contrasting the late twentieth century warming that is spread across many latitudes. Temperature scale is in degrees Celsius.

air temperature occurred between 1920 and 1930; much of the warming of the twentieth century at this site took place during that decade (Figure 2.8, Durre, 2001).

Effects of those sudden changes on oceanic ecosystems were widespread (Intergovernmental Panel on Climate Change, 2001a, Ch. 13, 16; Dickson, 2001). Northward dislocations of plankton, fish, mammals, and birds were extreme, including the range of economically important species, such as herring (Figure 2.9) and cod. In the case of cod, the warm ocean greatly extended their range, as larvae were transported north by the Irminger Current from the waters around Iceland. The northern catch rose from nothing to 300,000 tons per year in the 1950s and 1960s, then declined abruptly when the surface cold water reappeared in the late 1960s. Earlier successful cod fisheries were seen near Greenland in the 1820s and 1840s, but none had occurred in the wide interval between then and 1920. Such effects were also widespread in the Nordic and Barents Seas.

Causes of the abrupt yet long-lasting northern warming are not clear. The Arctic region as a whole seems to amplify climate variability, as we are witnessing now. Relationship with the AO/NAO mode is not clear; the indexes of the mode were strongly negative in 1910-1920 and strongly positive in 1920-1930, indicating enhanced cyclone activity in the subpolar Atlantic through the 1920s with southerly winds at the entrance to the Nordic Seas. Climate-model simulations (Delworth and Mann, 2000) suggested that the warming could have occurred with strong natural variability, through an increase in oceanic meridional overturning circulation, which brought warm, saline surface waters into the sub-Arctic and Arctic. In that

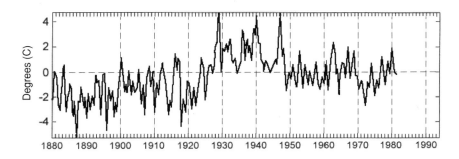

FIGURE 2.8   Surface air temperature anomalies at Upernavik, West Greenland (Adapted from I.M. Durre, personal communication, 2001).

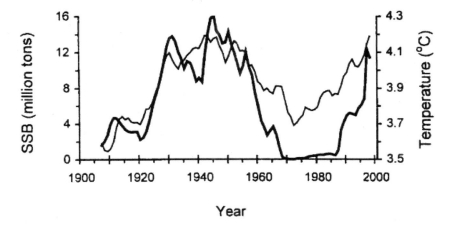

FIGURE 2.9 Spawning stock biomass (SSB) of herring in the Nordic Seas, responding to the abrupt warming of 1920s (Toresen and Ostvedt, 2000). SSB (bold line); temperature (thin line).

study, anthropogenic contributions to the earlier warming were also significant. The close proximity of the northern Atlantic focus of this warming to the dominant sinking regions of the global deep-ocean circulation makes this episode particularly relevant to global climate.

Abrupt Cooling and Freshening of the Subpolar North Atlantic, 1972-1996

Long-term measurements appropriate to climate variability studies are scarce in the ocean. Temperature records at a few coastal sites, and tide gauge records, have been sustained. Around 1948, a network of ocean weather stations was established by the International Civil Aviation Organization, mostly for atmospheric measurements. In the United States, these were manned by ships of the U.S. Coast Guard. A few, however—such as Bravo in the Labrador Sea, Mike in the Norwegian Sea, and Papa in the subpolar North Pacific—carried out deep hydrographic stations, as well as weather observations. Most were abandoned by the early 1970s; only one (Ocean Station Mike, at 65°N, 2°E in the Norwegian Sea; e.g., Gammelsrød et al., 1992) remains. The value of those time series is enormous.

Fortunately, for climate purposes, we do not always need rapidly sampled data. In the northwestern Atlantic the annual range of seasonal temperature is typically 4°C in the upper mixed layer, 1°C at 150m, and 0.1°C or less at 1000 m. Sampling must be frequent enough so as not to be aliased by this cycle. Monitoring deep, "stable" water masses is good strategy.

Combining weather ship data, hydrographic work, particularly that of the CSS Hudson from Bedford Institute of Oceanography, Canada, and more recent moored instrumentation reveals a picture of climate change over the last 30 years that stands out in the 100-year record. There has been a persistent decline in salinity from top to bottom of the subpolar Atlantic Ocean and widespread cooling since 1972 (Figure 2.10) (e.g., Lazier, 1995). Accompanied by a surge of strong, cold winters with winds from the Canadian Arctic (and strongly positive NAO index), convection in the Labrador Sea has penetrated to great depth, yet episodically, during this period. The intense period of Labrador Sea convection produced a great outflow of the water mass, which has been traced along the western boundary of the Atlantic as far as the Antilles (Molinari et al., 1998; Smethie et al., 2000).

The positive values of the AO/NAO-index time series in this period stand out in the 150-year record. Then, suddenly, the cold winters in the northwest Atlantic reverted to an extremely mild state in early 1996. The AO/NAO indices had suddenly reversed, taking on minimums as extreme as anything seen during the century. The ocean has responded quickly, with warm, saline water invading the Labrador Sea to replace the cold, low-salinity water characteristic of intense wintertime forcing. While possible feedback coupling of the atmosphere-ocean system remains to be determined, the direct correspondence of northwest Atlantic deep convection and cold-air outbreaks from Canada is not in doubt.

The increasing supply of fresh water leading to the strong decline in salinity has several possible origins: increased flow of water and ice from the Arctic, decreased flow of saline waters northward in the Gulf Stream/North Atlantic Current system from the subtropical Atlantic, and both the increased hydrologic cycle and ice melt associated with global warming. The resemblance to the freshening seen in coupled global climate models may be misleading, because of the inaccuracy of the AO/NAO modes in many of the models. The enhanced AO/NAO mode actually increased the density and intensity of this intermediate-depth branch of the THC during this period, despite the decline in salinity. Cooling of the subpolar ocean during a general period of global warming is a dynamic negative feedback,

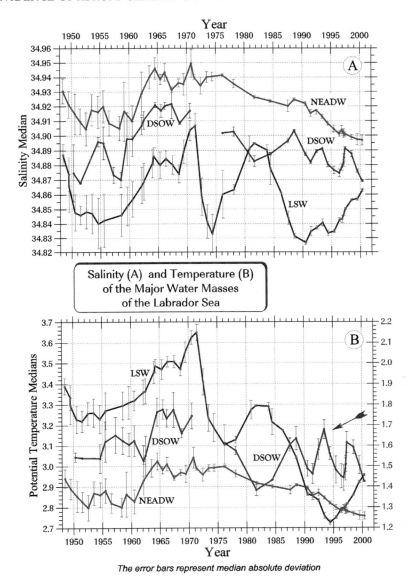

FIGURE 2.10   Variation of salinity (A) and potential temperature (B) with time, at three depths in the Labrador Sea. There is extensive cooling and salinity decline beginning abruptly in 1972, lasting for the last quarter of the 20th Century. The "LSW" (Labrador Sea Water) responds to winter weather locally, while the deeper "NEADW" (Northeast Atlantic Deep Water) and "DSOW" (Denmark Strait Overflow Water) cascade over the sills east of Greenland, and feed the deeper layers of the world ocean. (Courtesy of I. Yashayaev, Bedford Institute of Oceanography.)

which could delay by several decades the predicted slowing of the THC (e.g., Delworth and Mann, 2000).

Numerical simulations suggest that this enhanced production of intermediate-level water (which forms the upper North Atlantic Deep Water) is enough to energize, strongly and quickly, the entire THC of the Atlantic, and its closely correlated meridional heat transport (Häkkinen, 1999; Cheng, 2000). Model simulations allow us to make connections between observable and unobservable quantities, for example the sea-surface elevation field (seen by satellite altimeter), meridional heat transport and meridional volume flux (Häkkinen, submitted). These model simulations emphasize the fast-track response of the intermediate-depth THC during abrupt swings of climate, which may be quite different from changes in the deeper branch of the THC driven by overflows from farther north.

These changes during the instrumental era are in themselves large perturbations of the climate system. Still larger changes have been recorded in paleoclimate data, suggesting that complete shut-down of Labrador Sea deep convection, and its associated contribution to the THC, is possible. Hillaire-Marcel et al. (2001) inferred long periods without Labrador Sea convection during the previous interglacial period. Under a more heavily perturbed, warmer climate, we could see this state recur; learning to recognize its possible onset is a major goal of current research.

North Atlantic Surface Salinity Events

Important events embedded in the longer-term decadal variability of the Atlantic are the "great salinity anomalies," which seem to involve outpourings of low-salinity surface water and ice from the Arctic. One of the strongest developed during the late 1960s. The salinity of the upper subpolar Atlantic decreased during a series of mild winters at the same time that intense northerly winds occurred over the Fram Strait. It has been inferred that a great increase in Arctic ice and low-salinity water was driven into the Atlantic, although the influence is controversial. The combined effects of mild winters and buoyant surface layer all but halted deep convection in the Labrador and Irminger Seas. Low-salinity water was tracked as it progressed cyclonically around the high-latitude Atlantic (Dickson et al., 1988) and was seen moving into the Norwegian Sea in 1979; its influence was seen as far south as the Azores. "Lesser" salinity anomalies also appear in the instrumental record (Belkin et al., 1998). Finally, the long overall decrease in the salinity of the upper subpolar Atlantic since 1972 can similarly be attributed to the increasingly positive NAO events whose southerly winds near Europe tend to drive warm, saline Atlantic water into the Arctic (and, by inference, low salinity water back to the Atlantic) (Dickson et al., 2000). Those events provide models for the more intense salinity anomalies of the paleo-record, and perhaps of the near future. Changes in Arctic river run-off and its mixing into the Arctic Ocean may prove to be important (Dickson, 1999; Macdonald et al., 1999; Guay et al., 2001; Ekwurzel et al., 2001).

Arctic Change in the 1990s

The strong interaction between the Arctic and Atlantic Oceans was discussed above. During the recent period of widely varying climate in the subpolar Atlantic, the Arctic has also experienced great change. Exchange between Arctic and Atlantic is a new focus within the larger climate system. The increasing incidence of the positive phase of the AO/NAO appears to have driven warmer, more-saline Atlantic water into the Arctic and pushed back the boundary with fresher, colder waters of Pacific origin (Carmack et al., 1995, Morison et al., 1998). Warming of the polar Atlantic layer has been so striking as to be invoked in possible future melting of the sea ice cover. Attendant thinning of the sea-ice, by about 40 percent, has occurred in the central Arctic, based on comparison of data from 1993 to 1997 with those from 1958 to 1976 (Rothrock et al., 1999), but once again dynamics and greenhouse warming might be interacting or obscuring one another. Arctic ice-cover history is complex, yet full of strong climate signals (e.g., McLaren et al., 1990; Chapman and Walsh, 1993; Yueh and Kwok, 1998; Kwok et al., 1999). The relationship with AO/NAO wind forcing and temperature variability is potentially strong (Rigor et al., submitted). With positive polarity of the AO/NAO, weakening occurs of the typically high atmospheric pressure over the Beaufort Sea, together with intrusion of cyclonic flow related to the Icelandic low pressure center farther north into the Arctic. Moisture transport into the Arctic by this circulation has increased with the AO/NAO index (Dickson et al., 2000).

Antarctic and Southern Ocean Climate Change

Attention has focused on changes in the Arctic, but concern remains that this amounts to looking under the street light, missing other high-latitude sites that are less well observed. Southern Ocean overturning has two modes: through open ocean convection in polynyas (areas of open water that are surrounded by sea ice), in geographically suitable semi-enclosed seas (Ross and Weddell Seas), along the shelf of Australia, and driven by plumes of dense water descending from the shallow Antarctic shelf and slope, and interacting strongly with the overlying permanent ice. In contrast with the North Atlantic, the overturning that feeds the deepest waters is fed by cool mid-depth waters that first upwell near the Antarctic Circumpolar Current. Thus, the total heat transport associated with formation of dense waters in the Antarctic is not as dominant a part of the global heat budget as is the heat transport associated with the North Atlantic THC. However, variations in Antarctic overturning, and possibly surface warming, can affect freshwater budgets, as evidenced in large-scale freshening in the lower latitude oceans of the Southern Hemisphere (Wong et al., 1999; Johnson and Orsi, 1997). Both intermediate-depth and deeper branches of the THC form complex arteries of sinking and recirculating in the Southern Ocean (e.g., Orsi et al., 1999); both are important to the time-constants of response of the global ocean to abrupt change.

The possible alteration of northern and southern sinking in the global THC, known as the see-saw, arises from model simulations (Stocker et al., 1992; Crowley, 1992) and from comparison of climate anomalies in Greenland and Antarctic ice-core records (e.g., Blunier and Brook, 2001; see Plate 2). Instrumental data are scarce, but the inventory of CFC and other transient tracers can be interpreted to suggest a recent weakening of sinking around Antarctica, in comparison with the longer-term signal seen in steady-state tracers related to nutrients and dissolved oxygen (Broecker, 1999). A suggested link is proposed between this oceanic behavior and the Little Ice Age, one of the major climate events of the past 1,000 years in the northern hemisphere.

Identifying global connections in the climate system is a particularly important goal of modern observations. Toggweiler and Samuels (1995) have found that, using global ocean models, the THC is sensitive to forcing by westerly winds in the Southern Ocean (particularly those near the relatively narrow Drake Passage, south of South America). In their simulations the THC volume transport, as far north as the subpolar Atlantic, shifted in response to changes in this distant wind forcing.

In 1972, a large hole opened in the ice cover of the Weddell Sea. This polynya lasted until 1974, and its progress was tracked by the early Nimbus satellites. The polynya was supported by the breakdown of the usual stratification leading to open-ocean convection (Gordon and Comiso, 1988).

Elimination of the relative buoyancy of low-salinity upper waters allowed deep heat to be brought to the surface (so that the polynya can be self-sustaining; Martinson et al., 1981). The event left its mark on the ocean with anomalous cold, low-salinity deep water, which found its way into the deep circulation, as witnessed in the Argentine Basin farther north.

Global Ocean Heat Content

The archive of hydrographic data from throughout the world ocean gives us an instrumental record of climate change that integrates over time and space: the ocean dominates (by a factor of about 10^3) the heat capacity of the ocean-atmosphere system, and its storage of heat anomalies also greatly exceeds the land surface. Levitus et al. (1999) (also see Levitus et al., 2001 and Barnett et al., 2001) described the integrated ocean heat content from 1950 to 2000 (Figure 2.11). In every ocean, an overall warming during the last 50 years is evident. But in a hiatus 1978–1988, all the ocean basins other than the South Atlantic cooled substantially. The temporary cooling in Pacific and Indian Oceans (both north and south) amounted to about half the net rise (2×10^{23}J) over the 50 years. The abrupt cooling is mostly in the top 300 m of the water column. The anomalous air-sea heat flux during the event averaged about 1 watt m^{-2} which is several times as large as the half-century-long warming (0.31 watt m^{-2}). The cooling event fails to resemble the well-known dip in atmospheric surface temperatures between roughly 1940 and 1970 in timing or duration. Heat is mercurial, and global budgets like this can be surprisingly responsive to localized forcing. Possible contributors to the downward lurches are the eruptions of the volcanoes Agung in 1963, El Chichon in 1982, and Pinatubo in 1991 (Levitus, 2001).

Couplings

Instrumental records show preferred modes of behavior of the earth's climate system. Furthermore, those modes might be coupled not only in time (such as the unknown relation between the few-year ENSO and few-decade PDO variability), but also in space. Although the principal modes of variability at high latitude (the AO/NAO, and related southern hemisphere annular mode) have some independence from ENSO, connections have been proposed. For example, Hoerling et al. (2001) suggested, comparing observations and coupled-model simulations, that the average tropi-

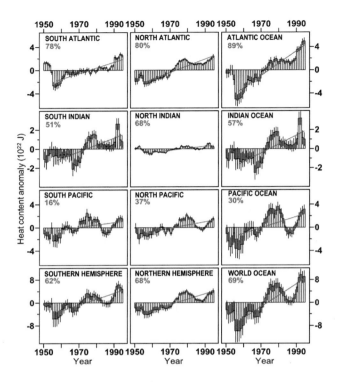

FIGURE 2.11   Time series for the period 1948 to 1998 of ocean heat content
(10^{22}J) in the upper 300 m for the Atlantic, Indian, Pacific, and world oceans. Note
that 1.5×10^{22}J equals 1 W·year·m^{-2} (averaged over the entire surface of earth).
Vertical lines through each yearly estimate represent ±1 standard error (SE) of the
estimate of heat content. (SOURCE: Levitus et al., 1999.)

cal warming since 1950 is a cause of enhanced AO/NAO activity in the
North Atlantic. This constitutes one pathway from anthropogenic forcing
to high latitudes, and another comes via the stratosphere. Intensification of
the stratospheric wintertime polar vortex under greenhouse-gas forcing has
been a major prediction of models (e.g., Shindell et al., 2001). The AO/
NAO is a mode of variability that grows more symmetric but no less ener-
getic as one moves upward from troposphere to stratosphere (Perlwitz and
Graf, 1995). Baldwin and Dunkerton (1999) demonstrated downward
phase propagation of intense stratospheric anomalies. The possible intensification of the AO/NAO beneath an enhanced polar vortex is being investigated with models and observations (Shindell et al., 2001). Yet, crucial

internal linkages, particularly the decadal shifts in Rossby wave propagation upward from the troposphere, still need to be established.

ENSO has great reach, and can influence climate at high southern latitudes. A potentially important climate mode, the Antarctic Circumpolar Wave, involves the coupled ocean and atmosphere near the polar front, in the Southern Ocean (White and Peterson, 1996). This wavenumber-2 dislocation of the polar front propagates eastward at about 0.08 m. sec^{-1}, thus taking 8 to 10 years to circumnavigate that ocean. The mode appears to be strongly excited by tropical ENSO events. At a much shorter timescale, blocking patterns in the strong westerly circulation in the southeast Pacific appear to be forced by ocean warming and divergence in the western tropical Pacific (Renwick and Revell, 1999). Couplings of this kind are achieved by Rossby-wave propagation.

The modal behavior of earth's climate is one of the major research results of the instrumental era. This review emphasizes these modes as well as describing recent examples of such regional changes as the Dust Bowl and North Atlantic oceanic conditions that were abrupt and impacted humans and ecosystems. The connection stems from the possibility that climate can lock into one phase or another of a modal oscillation; for example, preferring the warm equatorial Pacific phase of ENSO. Possible interactions among the major climate-system modes (especially ENSO and the high-latitude annular modes) suggest that changes in one could be propagated globally. The major perturbations associated with greenhouse-induced climate change may affect the likelihood of such changes, and their geographic extent.

SYNOPSIS OF OBSERVATIONS

A great wealth of evidence related to abrupt climate change has been collected, although important data gaps remain. Interpretation of sedimentary records to learn about past climate changes is improving rapidly. The unequivocal evidence of repeated large, widespread, abrupt climate changes in the past is striking. Such events were especially prominent during the cooling into and warming out of the most recent global ice age. However, the events are not restricted to cold times as shown by the cooling about 8,200 years ago, which punctuated conditions that in many places were similar to or even slightly warmer than recently. Other events during the current Holocene warm time may not have been as globally dramatic as the ice-age changes, but regional effects on water availability may have had large effects on humans and ecosystems.

Documentation of past abrupt climate changes is quite good in certain regions such as from Greenland ice cores and some of the pollen and other records summarized above. But no reliable global maps of climate anomalies are available for any of the abrupt changes of the past. Well-dated records with high time resolution are especially scarce in the Pacific, the tropics, and southern high latitudes, but no ideal record yet exists and additional useful information can be obtained almost everywhere. Past changes in the hydrological cycle are especially poorly characterized relative to its importance to humans and ecosystems.

The unavoidable incompleteness of the sedimentary record (spatially, temporally, and in variables recorded) means that changes from before the deployment of widespread instrumentation will never be understood as well as more recent changes. Instruments have not yet observed any of the large, global abrupt climate changes (a statement that some day may require modification in hindsight with respect to human-induced changes). However, regional events that satisfy the definition of abrupt climate change have been observed instrumentally. These are important in several ways: among others, they highlight observations required to characterize events and provide early warning of event onset and termination; they point to gaps in the observational system; and, they show the important role of coupled modes in such changes (and, by inference, perhaps also in the larger changes further in the past).

Instrumental study of abrupt changes has drawn on a broad-based range of observations collected for many purposes. However, few of these data sets are targeted on those parts of the climate system that are believed to have participated in past abrupt changes or that are likely to exhibit abrupt and persistent changes when thresholds in the climate system are crossed. Documentation remains sketchy of ocean circulation, especially in regions of deepwater formation, and interactions with sea-ice processes and with freshwater fluxes from the atmosphere and from land hydrology including glaciers and permafrost. Gaps also remain in documentation of land-surface processes linked to hydrology, and of modal behavior of the coupled climate system.

Much work is required to place all of the available observations into a coherent framework, capable of assimilating new observational data and contributing to an understanding of abrupt climate change. A start on such a framework is described in the next section.

Innovative instrumental methods have been developed that offer us hope of detecting abrupt climate change when it appears. These include both new *in situ* probes for atmosphere and ocean and remote sensing from satellites. New sensors (for chemical and biological fields as well as physical quantities) and new platforms (moored, drifting, gliding or profiling) are becoming available in large numbers. They provide efficient coverage globally and locally, at lower cost than classical observation methods. In reviewing the abrupt changes seen during the instrumental period, it is obvious how much more would now be understood if just a few well-placed modern instruments had been in place for an extended time.

For abrupt climate change, special emphasis on the dynamics of freshwater in both atmosphere and ocean is needed. Evaporation, precipitation, river flow, surface moisture, cryosphere dynamics, and related upper ocean dynamics are all relatively poorly observed, yet they are as important to climate as are the better observed thermodynamic fields. Cloud and water vapor dynamics in the atmosphere are crucial to climate, yet their representation in coupled numerical models is crude.

Models that are heavily involved in predicting or diagnosing abrupt climate change are known to be inaccurate in representing various high-latitude oceanic processes. Among them are deep convection, sinking and overflow dynamics, flow through narrow passages, over sills and in the descending branch of the oceanic overturning circulation, sea-ice dynamics, and flow and thermodynamics of shallow shelf regions at high latitude. Discrimination between sinking in the Labrador Sea and sinking farther north, with overflow at the Greenland-Scotland ridge system, is not well handled by such models.

For these reasons, sustained observations of the high-latitude ocean and its communication with the Arctic are needed. Hydrographic and chemical-tracer observations must be repeated, quasi-regularly, across important passages and major "stable" water masses. Direct observations of ocean circulation intensity using both *in situ* and satellite altimetric observations are particularly valuable for inferring meridional overturning variability.

Satellite scatterometer wind fields give us a component of global air/sea interaction with remarkable resolution. Passive radiometric observations give us a wealth of temperature, sea-surface structure and moisture data, and possibly new fields such as surface salinity. Synthetic aperture radar observations from satellites provide high resolution detail of ice fields, upper ocean flows, winds and terrestrial flooding and drainage-basin flows.

Tropical ocean observations near the sea-surface currently provide coverage of the pace of ENSO activity, and its connections with higher latitude. Fixed arrays, which have revealed so much about the Pacific, could be extended to Indian and Atlantic oceans, at least in skeletal form.

The ocean near the coasts is a particularly sensitive and biologically productive region. Considering the close proximity of so many humans, it is surprising how poor the instrumental record is. As natural variability is overlaid on human-induced changes of many kinds, there is a lack of baseline observations for comparison. Our definition of climate and its impacts is expanding to include ecosystems and human activity, and in this spirit there are many more observations to be made, particularly aiming at biological fields. Regular coastal observational programs are hampered by overlapping jurisdictions and variable commitment to basic scientific questions. River systems are intimately connected with estuaries, and human modification again compounds natural climate variability of streamflow, water quality, and drainage patterns. Support for sustained observations in both coastal regions and watersheds should be of high priority.

3

Processes That Cause Abrupt Climate Change

eather changes abruptly from day to day, and there is no basic difficulty in understanding such changes because they involve a "fast" and easily observed part of the climate system (e.g., clouds and precipitation). But mechanisms behind abrupt *climate* change must surmount a fundamental hurdle in that they must alter the working of a "slow" (i.e., persistent) component of the climate system (e.g., ocean fluxes) but must do so rapidly. Two key components of the climate system are oceans and land ice. In addition, the atmospheric response is a crucial ingredient in the mix of mechanisms that might lead to abrupt climate change because the atmosphere knits together the behavior of the other components. The atmosphere potentially also gives rise to threshold behavior in the system, whereby gradual changes in forcing yield nearly discontinuous changes in response.

A mechanism that might lead to abrupt climate change would need to have the following characteristics:

- A trigger or, alternatively, a chaotic perturbation, with either one causing a threshold crossing (something that initiates the event).
- An amplifier and globalizer to intensify and spread the influence of small or local changes.
- A source of persistence, allowing the altered climate state to last for up to centuries or millennia.

As discussed in Chapter 2, the Younger Dryas is the most studied example of an abrupt change and provides insights about possible mechanisms. In many ways the Younger Dryas serves as a defining event that embodies the notion of abrupt climate change. Heinrich and Dansgaard/Oeschger events are generally thought to be governed by physical laws that are closely related to those involved in the Younger Dryas; indeed, many consider the Younger Dryas to be just an unusually big Heinrich or Dansgaard/Oeschger event. However, there are potentially many types of abrupt change, as described in Chapter 2, so we need to consider the full array of possible processes, rather than only those currently in favor for explaining the Younger Dryas and Dansgaard/Oeschger events. Other major events of interest in the paleoclimatic records include decadal Holocene droughts, the dry-moist cycles of North Africa, the Little Ice Age, and the cooling event that occurred about 8,200 years ago (the "8.2K event"). Abrupt shifts in the dominant modes of the modern climate are also being documented, and the active mechanisms in these shifts might be relevant to larger abrupt changes in the distant past or in the future.

AN OVERVIEW OF MECHANISMS

Understanding and simulating abrupt climate change poses special challenges in climate science. Many climate changes are well described as relatively small deviations from a reference state, often assumed to be in equilibrium with external forcing. Powerful simplified conceptual approaches ("linearizations") are therefore appropriate and can explain many features in a faithful way. In a linear model, doubling the forcing doubles the response. The linear approach does not hold for abrupt climate change, in which a small forcing can cause a small change or a huge one, so fully nonlinear and transient considerations and simulations are required. In particular, transitions between qualitatively different climate states, as are seen in the paleoclimatic records, require abandoning the "time-slice" perspective in which equilibrium runs of climate models under different forcings are used to try to infer the path of climate change.

Three past classical model types have been important in this context because of their ability to simulate paleo abrupt climate change either spontaneously or in response to changes in controlling parameters or external forcing. These are the two-box Stommel model (1961), simple energy-balance models such as that of Sellers (1969), and the Lorenz models (1963, 1990).

Using a two-box model of the thermohaline circulation (THC) of the ocean, Stommel (1961) showed that the different response times of the ocean surface to heat and freshwater perturbations give rise to multiple equilibria of the THC with very different characteristics. This implies that forcing and other conditions do not uniquely define the state of the climate system and that perturbations beyond thresholds can trigger transitions to other equilibrium states. The model presents some fundamental concepts and demonstrates the major sources of uncertainty in simulating abrupt climate change caused by a change in the THC. Multiple equilibria, however, would not occur if all processes were linear. The nonlinearity of the Stommel model arises because the flow field of ocean water transporting heat and salt is itself a function of temperature and salinity. Nonlinearities and multiple equilibria are the fundamental concepts behind the simulation of abrupt climate change.

Simple models of the energy balance of the atmosphere exhibit multiple equilibria. The model of Sellers (1969) is a typical case: a given amount of solar irradiation allows either a cold or a warm planet in equilibrium. The nonlinearity here is introduced by a special formulation of the snow-albedo feedback, which is operative only in a particular range of temperatures. A cold earth is snowy, and snow reflects sunlight and keeps the earth cold. A warmer earth has less snow, absorbs more sunlight, and so stays warmer. Abrupt change can be triggered by variations in total solar output or other parameters that influence the radiative balance, such as snow cover.

The Lorenz models (Lorenz, 1963, 1990) provide a highly simplified description of atmospheric circulation. In addition to multiple equilibria in particular ranges of parameters, these models exhibit self-sustained oscillations and chaotic behavior. A system oscillates near one of two preferred centers; abrupt change occurs when the system switches from one mean of oscillation to the other. In a Lorenz model, these transitions occur spontaneously.

These examples suggest that abrupt climate change can occur in (at least) three fundamentally different ways:

• Abrupt climate change can be the response to a rapidly varying external parameter or forcing. If one views only the atmosphere-ocean system, massive sudden discharges of freshwater from disintegrating ice sheets on land would be an example of a sudden external influence. Nonlinearity in the atmosphere-ocean system is not a prerequisite for such behavior, whose time scale is dictated essentially by that of the forcing.[1]

• Slow changes in forcing can induce the crossing of a threshold and result in the transition to a second equilibrium of the system. The evolution of such a change would be governed by the system dynamics rather than by the external time scale of the slow change. In considering the whole earth system rather than just the oceans and atmosphere, massive discharges of freshwater from disintegrating ice sheets would be a result of threshold-crossing. Slow melting at the end of the last ice age produced ice-marginal lakes. When the ice margin reached a particular location, such as the path of a former river that the ice had dammed, a threshold was crossed, the ice dam broke, and the water was released rapidly (Broecker et al., 1988).

• Regime transitions can occur spontaneously in a chaotic system. In this case, external triggers for transitions are not required, so a series of regime changes could continue indefinitely or until slow changes in external forcing or system dynamics removed the chaotic behavior.

Oceans

Changes in ocean circulation, and especially THC in the North Atlantic, have been implicated in abrupt climate change of the past, such as the Younger Dryas and the Dansgaard/Oeschger and Heinrich/Bond oscillations (Broecker et al., 1988; Alley and Clark, 1999; Stocker, 2000). Today, relatively warm waters reach high latitudes only in the North Atlantic. The high salinity of the Atlantic waters allows them to sink into the deep ocean when they cool, and warmer waters flowing along the surface then replace them. This yields a net heat transport into the high northern latitudes of the Atlantic and northward heat transport throughout the South Atlantic, carrying heat into the North Atlantic (Ganachaud and Wunsch, 2000; see also Plate 4b.)

Outburst floods, which would have freshened the North Atlantic and reduced the ability of its waters to sink, immediately preceded the coolings of the Younger Dryas and the short cold event about 8,200 years ago (Broecker et al., 1988; Barber et al., 1999); this suggests causation. Evidence of reduction or elimination of northern sinking of waters during cold times (Sarnthein et al., 1994; Boyle, 2000) provides further support, as does

---

[1]For the whole earth system, a massive nuclear war or the impact of a gigantic meteorite might be considered in this category; such mega-disasters are not the focus of this report because the research agenda is not primarily concerned with the climate system.

the see-saw relation between Greenland and Antarctic temperatures on millennial scales (Blunier and Brook, 2001; see also Plate 2), which suggests that reduction in heat transport to the north allowed that heat to remain in the south.

Those and other considerations focus attention on changes in the THC as one cause of abrupt climate change. However, additional processes presumably were active in the past abrupt changes exemplified by the Younger Dryas, as indicated by the difficulty of fully explaining the paleoclimatic data on the basis of the single mechanism of North Atlantic THC changes. Therefore, the ocean's role in climate is developed more fully in the following.

Water has enormous heat capacity—oceans typically store 10-100 times more heat than equivalent land surfaces over seasonal time scales, and the solar input to the ocean surface for a year would warm the upper kilometer only 1 degree—so the oceans exert a profound influence on climate through their ability to transport heat from one location to another and their ability to sequester heat away from the surface. The deep ocean is a worldwide repository of extremely cold water from the polar regions. If much of this water were brought to the surface in temperate or tropical regions, it could cause substantial cooling that, although transient, could last for centuries. It is not easy to bring cold water to the surface against a stable gradient, though, and this can happen only in special circumstances. Such localized change could, however, have a wider impact through atmospheric teleconnections. Fluctuations in ocean heat transport can also affect climate; for example, an increase in equator-to-pole heat transport would warm the polar regions (melting ice) and cool the tropics.

The implications of fluctuation in heat transport by the Atlantic THC have received particular attention, especially as a mediator of Younger Dryas and Dansgaard/Oeschger abrupt change. Deep water forms only in the North Atlantic and around the periphery of Antarctica, where extremely cold, dense waters occur. There is no deep-water formation in the North Pacific, because the salinity is too low to allow high enough density to drive deep convection, despite the low temperatures. By analogy, change in the freshwater balance of the North Atlantic, which might be caused by glacial discharge or warming of the planet through increases in carbon dioxide, potentially can act as a trigger to turn the THC on or off. In contrast, it is not thought that future climate change could turn on deep-water formation in the North Pacific, although there is evidence that at times during ice ages the intermediate waters of the Northern Hemisphere Pacific were more ventilated than they are now (Behl and Kennett, 1996).

The threshold for deep-water formation cannot be thought of apart from the general circulation of the world ocean, because the density required for North Atlantic surface waters to sink is determined in relation to the "prevailing" deep-water density of the rest of the ocean. This density is determined globally and is intimately linked to processes in the Southern Ocean. Furthermore, the density of surface waters in the North Atlantic is not determined by purely local processes, in as much as the North Atlantic salinity is affected by mixing with transported subtropical Atlantic waters, whose salinity in turn is affected by tropical winds, which can systematically transport moisture out of the Atlantic basin. The freshwater balance of the Atlantic is further affected by melting of glaciers, transport of freshwater by sea ice, and land-surface processes that determine runoff patterns.

Several ocean heat-transport mechanisms other than the THC are in operation today. In particular, wind-driven ocean circulation dominates ocean heat transport in the North Pacific (Bryden et al., 1991) and the Indian Ocean (Lee and Marotzke, 1997). Results from relatively simple models suggest that in the absence of a vigorous THC, wind-driven heat and salt transports increase, making up at least in part for the loss of the THC as a transport agent (Marotzke, 1990; Winton and Sarachik, 1993).

Cryosphere

Land glaciers and sea ice enter into abrupt change mechanisms in many ways. The accumulation of ice on land and the associated ice-albedo feedback are probably too slow to be involved in abrupt climate change. However, because a glacier that is frozen to its substrate can surge if the basal temperature of the ice is raised to the melting point, glacial discharge and decomposition can be rapid (MacAyeal, 1993a,b; Alley and MacAyeal, 1994). Ice-sheet surging certainly would affect sea level, as noted in Chapter 4 with regard to the West Antarctic ice sheet. Surging also may affect the atmospheric flow pattern by changing the elevation of parts of the continental ice sheets (Roe and Lindzen, 2001). Furthermore, rapid glacial discharge can release armadas of icebergs into the ocean, which serve as an important indicator of abrupt climate change; increases in ice-rafted debris are the defining feature of Heinrich events (Broecker, 1994). More importantly, glacial discharge abruptly increases the delivery of freshwater to the ocean (to the North Atlantic, in the case of Heinrich events), with the potential of modifying the THC. Another class of catastrophic event associated with land glaciers is the formation of large lakes of meltwater, held back only by fragile ice dams. The breaking of an ice dam can lead to the sudden delivery of massive amounts of freshwater to the ocean. As noted above, it is believed that the draining of an ice-dammed lake (Agassiz) was at least partly involved in the initiation of the Younger Dryas event and was most likely responsible for the event about 8,200 years ago (Barber et al., 1999; Broecker et al., 1988; Teller, 1990; Teller, in press).

Sea ice, which forms by freezing of ocean water, is an important amplifier of climate forcing. When sea ice forms, it increases the planetary albedo, enhancing cooling. Sea ice also insulates the atmosphere from the relatively warm ocean, allowing winter air temperatures to decline precipitously (as much as tens of degrees C, compared to conditions over open water), and reducing the supply of moisture to the atmosphere, which in turn reduces precipitation downwind. The rectifying and amplifying effects of sea ice are important in connection with changes in the location of North Atlantic deep-water formation. The site of North Atlantic deep-water formation is roughly where a substantial fraction of the heat transported northward in the Atlantic Ocean is deposited. Changes in the location affect the sea-ice margin and can have a net effect on the planetary radiation budget.

Formation of sea ice also leads to rejection of very dense brine. This is particularly important around the Antarctic margin. Brine formation there is the major contributor to formation of the world's deepest ocean water.

Sea ice must be considered dynamically. Its movements are rather like those of a viscous fluid in coexistence with ocean water when viewed over a large enough area, but with brittle behavior in smaller regions. The creation of leads, or cracks, in sea ice affects albedo and air-sea exchange, and the movement of sea ice from one place to another has important effects on the global distribution of ice cover. Similar transport issues arise with regard to transport of icebergs discharged from land glaciers; here, the main interest concerns the distribution of freshwater ultimately delivered by the icebergs.

Snow cover can also serve as an amplifier of climate change and a source of persistence. Snow-covered land maintains cold conditions because of its high reflectivity and because its surface temperature cannot rise above freezing until the snow melts. There are interesting interactions between snow cover and vegetation. A modest snow cover on a flat surface, such as tundra, suffices to cause high albedo. However, in terrain covered with evergreen trees, snow falls to the surface without completely coating the dark canopy, allowing absorption of much solar radiation.

Atmosphere

The atmosphere is involved in virtually every physical process of potential importance to abrupt climate change. The atmosphere provides a means of rapidly propagating the influence of any climate forcing from one part of the globe to another. Atmospheric temperature, humidity, cloudiness, and wind fields determine the energy fluxes into the top of the ocean, and the wind fields dictate both the wind-driven ocean circulation and the upwelling pattern (of particular interest in the tropics and around Antarctica). Atmospheric moisture transport helps govern the freshwater balance, which plays a crucial role in the THC, and precipitation patterns provide the hydrological driving of glacial dynamics. The atmospheric response to tropical sea-surface temperature patterns closes the feedback loop that makes El Niño operate. Atmospheric dust transport can affect the planet's radiation balance and on longer time scales might affect ocean carbon dioxide uptake via iron fertilization (Mahowald et al., 1999).

How do oceans affect the air-temperature pattern? The primary—although by no means the only—effect of oceans on air temperature derives simply from the heat capacity of the oceans and has little to do with horizontal ocean heat transport. Being composed of a fluid that can mix heat vertically, oceans are slow to cool in winter and slow to warm in summer. Land temperature, in contrast, responds rapidly to adjust to changes in the seasonal cycle of solar radiation. The primary winter pattern of surface air temperature thus consists of warm oceans and cold land (see Plate 5). To some extent, that contrast depends indirectly on ocean dynamics, which affect the upper ocean stratification and so the mixed-layer depth and the amount of seasonal heat storage (Plate 5).

Complete shutdown of the THC would remove about 8 W/m^2 from the Northern Hemisphere extratropical heat budget (Pierrehumbert, 2000). To restore balance, the northern atmosphere-ocean system must cool down until the infrared radiation lost to space is correspondingly reduced. On the basis of a conventional sensitivity factor incorporating water-vapor feedback, the perturbation in heat budget implies an extratropical cooling of about 4°C, and growth of sea ice could amplify this cooling. This is roughly the cooling found in simulations (Seager et al., 2001), in which ocean heat transport was suppressed. Because the Northern Hemisphere THC is primarily an Atlantic rather than Pacific phenomenon, the heat transport is focused in the Atlantic. The somewhat greater warmth of European coastal regions than of similar latitudes on the Alaskan coast might well be linked to this THC heat transport (Plate 5), although debate continues about where and by how much ocean heat transport warms the atmosphere, and the extent to which changes in oceanic heat transport would be balanced by changes in atmospheric heat transport.

Popular treatments sometimes imply that the THC is responsible not only for the temperature difference between Europe and Alaska, but also for the larger difference between Europe and the east coast of North America. This is unlikely because the US Pacific Northwest coast is warmer than the east coast of Asia without an active THC in the North Pacific. Two factors appear to provide the large differences between east and west coasts. First, the prevailing westerly winds pick up some heat during winter while blowing over the ocean, spreading the moderating influence of the ocean downwind. Second, and perhaps more important, the planetary wave pattern—the global-scale sinuous bending of the jet stream—brings Arctic air down over the US New England region and the Asian coast of the Pacific and warmer breezes to the US Pacific Northwest and Europe. The pattern is known to be driven primarily by mountains in the winter (Nigam et al., 1988), and is influenced relatively little by land-sea temperature contrasts. An improved quantitative understanding of the east-west (North America to Greenland to Europe) gradient of temperature fluctuations in past abrupt climate changes is crucial to furthering the understanding of mechanisms. On the other hand, the relative warmth of Norway, which is just to the east of the second major center of heat loss in the North Atlantic (the first being in the Gulf Stream off the coast of the United States), is probably due to the THC; the Pacific lacks similar warmth to that off Norway.

A major problem with THC-based theories of the Younger Dryas and Dansgaard/Oeschger events is that atmospheric models yield a predominantly local North Atlantic cooling in response to THC shutdown, with few of the global repercussions that seem to be demanded by the data (see Chapter 2). In fact, the atmosphere alone does not seem able to extend the influence of extratropical influences to the rest of the globe efficiently. Even such a major forcing as the introduction of Northern Hemisphere ice sheets of the last glacial maximum into models has little influence south of the equator (Manabe and Broccoli, 1985; Broccoli and Manabe, 1987). In models generally, North Atlantic cooling affects the strength of the tropical

Hadley circulation, which leads to some temperature change in the northern half of the tropics, and more important precipitation changes.

There have been a number of simulations of the atmospheric response to THC shutdown or to directly imposed North Atlantic sea-surface temperature perturbations. Fawcett et al. (1997) eliminated Nordic sea oceanic heat transport in the GENESIS model and found localized reductions in surface air temperature of as much as 24°C. Even with such an extreme cooling, the temperature perturbation is very localized. It amounts to only 2.8°C over the summit of Greenland versus about 8°C observed (Severinghaus et al., 1998) and 2°C or less over much of Europe, underestimating some changes based on proxy records. As summarized in Ágústsdóttir et al. (1999) and Fawcett et al. (1997), these experiments matched many aspects of changes reconstructed from proxy records (including changes in seasonality of precipitation in central Greenland and other such details) but generally underestimated the magnitude of reconstructed changes except close to the North Atlantic. It is difficult to determine how the oceanic heat transport involved in these experiments compares with the heat carried by the real THC.

Manabe and Stouffer (1988, 2000) carried out experiments with a coupled atmosphere-ocean model in which the THC was largely suppressed by a massive artificial injection of freshwater. They found a more moderate Atlantic surface cooling of 6°C with cooling of 1-4°C over Europe and the greatest coolings confined primarily to coastal areas. The atmospheric-response experiments of Hostetler et al. (1999), forced by imposed sea-surface temperature patterns, yield a quite different impression, as the North Atlantic cooling leads to temperature reductions in excess of 4°C over all of Europe and much of Asia. The general locality of atmospheric response in those simulations does not by any means disprove the THC theory of the Younger Dryas and Dansgaard/Oeschger events. Current atmospheric models might be missing some crucial physical feedbacks that allow the real atmosphere to exhibit such large and widespread responses to THC changes. This is an unsettling possibility, in that it suggests that models could also fail to anticipate the threat of surprising and abrupt changes, which might occur in connection with global warming, as indicated in Plate 7 and discussed extensively in Chapter 4.

Ocean dynamics could help to extend northern extratropical influences to the tropics and to the Southern Hemisphere, and indeed ocean circulation experiments show some global changes in response to North Atlantic freshwater pulses. A full treatment of the role of the oceans must be carried

PLATE 1 Analyses of such ice cores reveal features of the climate when the ice was deposited. In this photo, Geoff Hargreaves, curator, stores a sample of GISP2 deep ice core from central Greenland in the main archive of the National Ice Core Laboratory (a joint effort of the National Science Foundation and the United States Geological Survey with the University of New Hampshire as academic partner, at the Denver Federal Center).

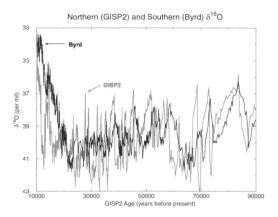

PLATE 2 Temperature changes are reflected in changes in the oxygen-isotopic ($\delta^{18}O$) ratios, with low $\delta^{18}O$ indicating high temperature. Records of $\delta^{18}O$ from ice cores in central Greenland (GISP2) and West Antarctica (Byrd Station) are shown, synchronized to the GISP2 time scale by Blunier and Brook (2001). These records show that both hemispheres experienced an ice age at similar time, that millennial oscillations superimposed on the ice-age cycle were especially large during the slide into and climb out of the ice age, that millennial oscillations were of larger amplitude in the north than in the south, and that at many times (e.g., around 70,000 years ago) antiphase behavior is exhibited between north and south in these millennial oscillations. Note that in this figure, time increases toward the left.

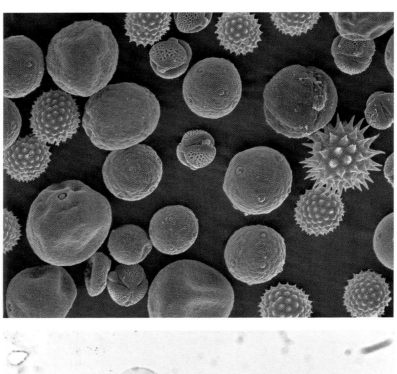

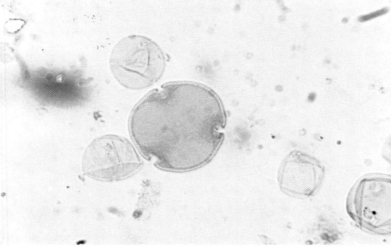

PLATE 3 The top photo is a false-color scanning electron micrograph of assorted pollen grains showing how the size, shape, and surface characteristics differ from one species to another. By counting the number of species and their abundance in sediments and peat deposits (called "palynology"), it is possible to analyze how ecosystems and climate have varied over time. The bottom photo shows pollen from Tilia (Linden or Basswood), which is a temperate deciduous species.

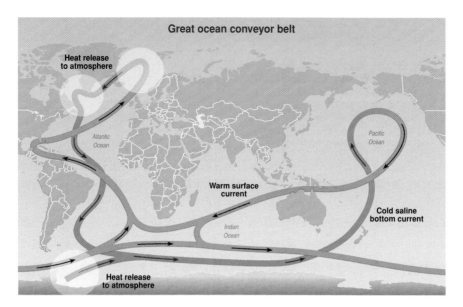

PLATE 4a The great ocean conveyor belt. This is a schematic generally summariz-
ing some important features of the world's ocean circulation. Warm, low-salinity
water, flows north along the surface of the Atlantic, becoming saltier (red arrows).
Cooling of this to saltier water in the North Atlantic produces high enough densities
for the water to sink and flow southward in the deep ocean and into other ocean
basins (blue arrows) (after Broecker, 1995).

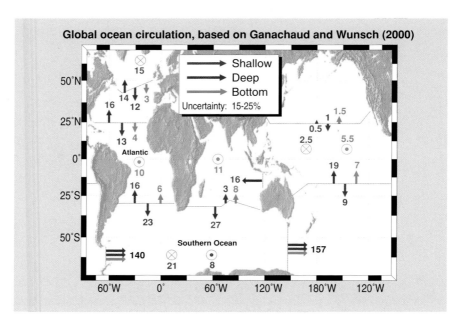

PLATE 4b A slightly more complex representation of the global ocean circulation than in Plate 4a, simplified from Ganachaud and Wunsch (2000), as estimated from modern oceanographic data. The figure shows the integrated flow across the latitudes where observations were taken during the World Ocean Circulation Experiment (WOCE) in the 1990s. The red arrows designate near-surface flow (typically warm; technically, water density less than 1027.72 kg m^{-3}), blue and green arrows are deep and bottom flows, respectively. Units are Sverdrups (million cubic meters per second); for comparison, the Gulf Stream transports around 31 Sverdrups northward through the Florida Strait. Notice the vigorous sinking in the North Atlantic and the near-complete absence of sinking in the North Pacific.

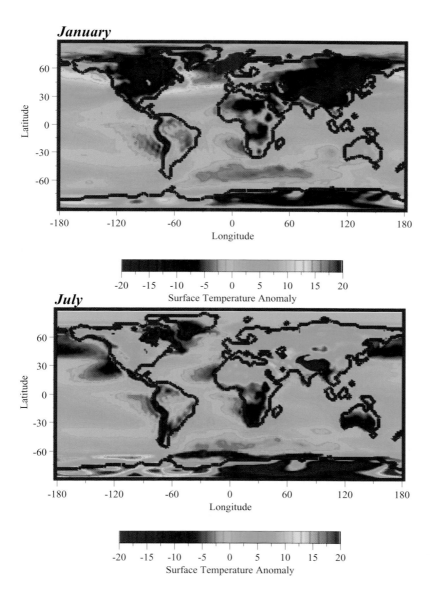

PLATE 5 January and July surface air temperature anomalies in degrees Celsius averaged over the years 1970-1980. The "anomaly" is defined as the deviation of air temperature from the average air temperature along a latitude circle passing through the point in question. Removing the mean makes it easier to see East-West variations in temperature without being distracted by the much larger North-South variations.

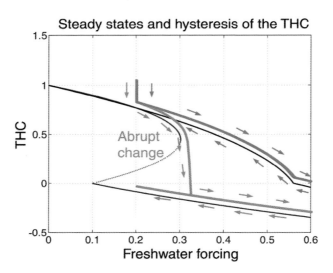

Steady states and hysteresis of the THC

PLATE 6 Paths along the solution curves of two versions of Stommel's box model showing the rate of the ocean overturning when the freshwater forcing flux H is increased and then decreased. Only in the case of weak diffusion (orange) does the model respond with an abrupt change, once a threshold in H is crossed. In the case of strong diffusion (green), at any time, there is a unique equilibrium.

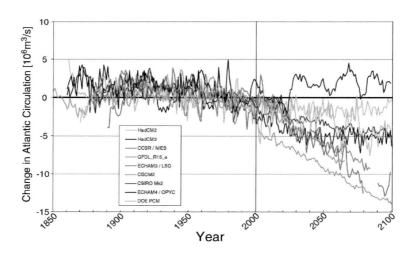

PLATE 7 Changes in Atlantic circulation in different models forced by prescribed increases in atmospheric CO_2. Most models show a reduction of the THC in response to increasing greenhouse gas forcing (Intergovernmental Panel on Climate Change, 2001b).

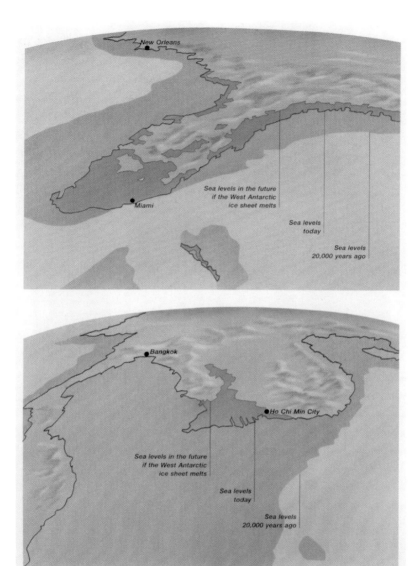

PLATE 8 During the last ice age, sea levels were approximately 100 m lower than present because more water was contained in ice sheets, and thus not available to the oceans. The dark blue shading around Florida and Southeast Asia demarcates areas that, while 20,000 years ago were dry land, are now under water. Sea levels would rise and flood coastal regions, to the approximate levels shown in the figure (dark green area), if the West Antarctic ice sheet or much of the Greenland ice sheet were to melt. The black line shows the present coast. SOURCE: Burroughs, 1999.

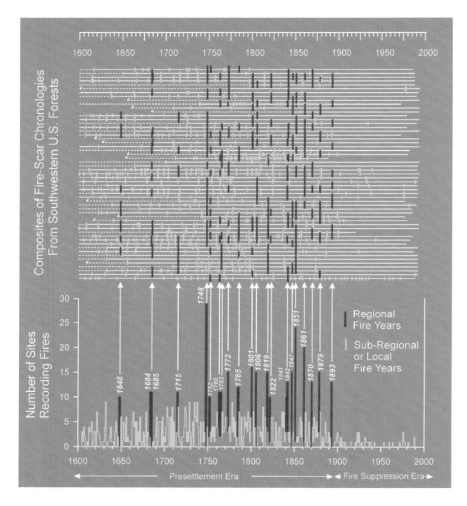

PLATE 9 Composites of fire-scar chronologies for 55 forest stands in Arizona, New Mexico, and northern Sonora, Mexico from the years 1600 to 2000, as reconstructed from tree-ring analyses of fire-scarred trees. The red and yellow vertical tick marks indicate the fire dates recorded by the fire-scarred trees sampled in each stand. The red tick marks represent regional fire years that were recorded by fire scarred trees in 10 or more of the sampled stands. Many of these regional fire years occurred during droughts that were probably associated with La Niña events. Note the very striking decrease in sites recording fires more recently than circa 1890, coinciding with the introduction of large numbers of livestock and organized suppression of fires by government agencies (Swetnam et al., 1999). Major changes in anthropogenic land use affect the frequency of fires associated with droughts.

out with coupled atmosphere-ocean models. There have not yet been any successful simulations of the pattern and magnitude of the Younger Dryas or of recurrent Dansgaard/Oeschger events with coupled ocean-atmosphere general circulation models. And there have not been any integrations of coupled ocean-atmosphere models over sufficiently long time scales to determine the natural level of millennial variability in such models, particularly in glacial conditions. Despite considerable success in modeling some aspects of past abrupt climate changes, there is much work still to be done.

Fluctuations in the atmospheric carbon dioxide concentration (Petit et al., 1999) have played a crucial role in climate change on the glacial-interglacial time scale (Pollard and Thompson, 1997). According to general circulation model results (Manabe and Broccoli, 1985), without the reduced greenhouse effect arising from low carbon dioxide in glacial times, the Southern Hemisphere would have experienced little cooling despite the massive growth of Northern Hemisphere ice sheets. However, carbon dioxide fluctuations are not radiatively significant on the time scale of the Younger Dryas or Dansgaard/Oeschger events, and up to industrial times carbon dioxide had only insignificant fluctuations throughout the Holocene. Methane is an important greenhouse gas and did fluctuate markedly in the course of many abrupt-change events, with changes in low-latitude and high-latitude sources (Brook et al., 1999). The low-latitude source changes are generally thought to be indicative of changes in tropical hydrology, so methane serves as an important indicator of the involvement of the tropics in abrupt climate change (Chappellaz et al., 1997). However, the magnitudes of methane changes were too small to yield appreciable radiative forcings, and the observation that temperature changes in Greenland preceded methane changes (Severinghaus et al., 1998) is not generally consonant with driving by a methane greenhouse effect.

Water vapor is an important greenhouse gas, but it is different from carbon dioxide and methane in that its concentration is determined primarily by atmospheric temperature and circulation patterns rather than by sources and sinks (Held and Soden, 2000; Pierrehumbert, 1999). Most of the atmosphere is highly undersaturated in water vapor and thus could hold considerably more than it does at present. The undersaturation is particularly prominent in the tropics, so reorganizations of the atmospheric water-vapor distribution stand as a possibility for amplifying and globalizing abrupt climate change. There are intriguing possibilities for interaction between water-vapor feedback and dust fluctuations, inasmuch as dust and other aerosols can serve as cloud condensation nuclei (Durkee et al., 2000).

By affecting precipitation efficiency, that process can potentially alter the water vapor content of the atmosphere. In addition, dust that absorbs solar radiation can affect precipitation through the circulation that arises in response to such heating.

Mechanisms centered in the tropics, particularly the tropical Pacific, have received increasing attention in recent years. The tropics have a compelling advantage over the North Atlantic as a mediator of global climate change. Because of the relatively weak confining influence of the earth's rotation at low latitudes, the effect of local changes in sea-surface temperature is communicated almost instantly through the atmosphere to the entire tropical band. This confining influence is due to the earth's rotation, which causes the well-known Coriolis effect in which all atmospheric and oceanic flows turn relative to the surface beneath them (to the right in the Northern Hemisphere and to the left in the Southern Hemisphere) rather than proceeding directly from regions of high to low pressure. The degree of turning is proportional to the sine of the latitude, so turning disappears at the equator. The strong high-latitude tendency for wind and ocean currents to turn slows the transmission of information through the climate system, whereas the weaker turning tendency near the equator allows rapid communication.

Tropical forcings create circulation patterns that have a major remote impact in the middle-latitude and polar regions, again communicated through the atmosphere. The tropical atmosphere-ocean system offers a rich palette of possible amplifiers and switches that could in principle lead to abrupt climate change. The dominant atmospheric circulation in the tropics is an overturning motion known as the Hadley cell, with air rising in warmer regions near the equator, spreading at high altitude, sinking in the subtropics, and returning along the surface. This circulation has a profound effect on tropical water vapor, cloudiness, and convection, with rain forests under the rising limb and deserts under the descending air. The Coriolis turning of the return flow along the surface yields the surface easterlies that drive the ocean. The upper branch of the circulation affects middle-latitudes through its influence on the upper-level subtropical jet.

Shifts in the position of the rising branch (the intertropical convergence zone, or ITCZ) can lead to major changes in the strength of the circulation (Hou and Lindzen, 1992; Lindzen and Hou, 1988), and it has been suggested that ITCZ shifts could amplify abrupt climate change (Clement et al., 2000). Model results suggest that changes in North Atlantic temperature associated with cutoff of the THC cause substantial changes in the Hadley circulation, propagating the influence of THC shutdown into the tropics (Manabe and Stouffer, 1988; Fawcett et al., 1997) and causing changes in tropical precipitation. In simulations, imposed North Atlantic cooling causes enhanced wind-driven oceanic upwelling in tropical and extratropical regions, which would bring colder waters to the surface and thus might contribute to additional cooling (Ágústsdóttir et al., 1999). There are many possibilities for regime switches lurking in the collective behavior resulting from coupling the Hadley cell to ocean dynamics.

El Niño is an oscillation of the coupled tropical atmosphere-ocean system. The influence of El Niño extends strongly into the extratropics. How much would El Niño change in a warmer or colder climate? Interest in that question has sharpened because there are indications that the character of El Niño events underwent a shift beginning in the 1970s. Carbon-14 data from corals suggest that changes in upwelling and in the source of subsurface water were involved in the shift (Guilderson and Schrag, 1998). Furthermore, comparisons with prehistoric El Niño records recovered from corals from the Last Glacial Maximum and from the previous major interglacial suggest a systematic relation between global conditions and the temporal character and amplitude of El Niño (Hughen et al., 1999; Tudhope et al., 2001).

Changes in El Niño are important in themselves because El Niño is by far the largest interannual climate signal at present and in the past, but it is possible that such changes might also mediate widespread changes in the global climate regime. Tropical transient motions, including those associated with El Niño, affect the water vapor and cloud distribution and hence the global energy budget (Pierrehumbert and Roca, 1998; Pierrehumbert, 1999). Tropical sea-surface temperature fluctuations, in contrast with those in the middle latitudes, cause variations in deep atmospheric heating, in turn giving rise to waves that powerfully communicate their influence to the rest of the planet. Any change in one part of the tropics tends to drag along the temperature of the entire tropical free troposphere owing to the tight coupling enforced by the Hadley and Walker circulations. The tropics are thus a natural candidate to be a "globalizer" of climate influences. The El Niño cycle might also affect transport of freshwater between the Atlantic and Pacific basins through its influence on low-level wind patterns (Latif et al., 2000; Schmittner et al., 2000). In addition, wind shifts affect the patterns of tropical upwelling and subtropical ocean gyres, possibly leading to changes in heat transport out of the tropical oceans. A treatment of some of the factors governing heat transport within the tropics and subtropics can be found in Lu et al., (1998), Lu and McCreary (1995), and McCreary and Lu (1994).

Low clouds have an albedo that is not very different from that of sea ice, but they can form almost instantaneously and are very sensitive to the structure of the tropical boundary layer. Increase in low cloud cover has a cooling effect, whereas dissipation of low clouds has a warming effect. The occurrence of low clouds is not solely, or even primarily, a function of temperature. Cloud properties are influenced by dust and aerosols, and an increase in these can make it easier for water to condense into small droplets, yielding brighter clouds and affecting precipitation. All these effects are subject to considerable uncertainties, and they remain as possibilities for mediating abrupt climate change or amplifying effects of THC fluctuations.

Land Surface

Land-surface processes can participate in abrupt climate change in many ways. The albedo of the land surface can change greatly, with fresh snow or ice sheets reflecting more than 90 percent of the sunlight striking them but dense forests absorbing more than 90 percent. Changes in surface type thus can affect solar heating and feed back strongly on climate. Rainfall used by plants and transpired to the atmosphere contributes substantially to local cooling during evaporation, and it supplies clouds and additional rainfall. Rainfall not used by plants typically runs off in rivers to the ocean, freshening surface waters at outflows but leaving low humidity in air over land. Thus, changes in vegetation have effects well beyond localities where the changes occur.

The land surface is the major source of dust, smoke and soot, and a variety of biogenic emissions. These affect cloud formation and albedo, drop size and rainout, and clear-sky radiation. Again, changes in the land surface can feed back on climate. The roles of these and other land-surface processes in causing, amplifying, or allowing the persistence of abrupt climate change are at best poorly understood. Additional work, especially in land hydrology and dust-cloud processes, is warranted.

External Forcings

A few types of forcings external to the climate system could play roles as pacemakers of abrupt climate change. These forcings vary too slowly to be prime movers of abrupt change, but if the climate system exhibits discontinuous response to continuous variations of some forcing parameters, then the possibility exists that external forcing variations might determine the timing of events.

For example, the earth's orbital parameters vary over time, affecting the distribution of solar energy delivered to the planet in time and location. The fastest of these variations is the precessional cycle, which takes about 22,000 years to complete and determines when the summer solstice occurs relative to the time of closest approach of the earth to the sun. In the tropics, that yields an 11,000 year half-precessional rhythm, and there are substantial changes in the insolation pattern over as little as 5,000 years. Those changes could have a major impact in the tropics through their effect on the latitudinal excursion of the ITCZ and its consequent effects on the Hadley circulation. The effect of precessional insolation changes on monsoonal circulations has been implicated in the moistening and drying of the Sahara, with strong feedbacks linked to land-surface processes, including changes in vegetation (Kutzbach et al., 1996; Claussen et al., 1999; Carrington et al., 2001). Precessional insolation changes possibly could have large effects on El Niño occurrence (Clement et al., 2000, 2001).

The effects and magnitude of fluctuations in solar output are less well constrained. There is an observable modulation of the sun's brightness over the 11-year solar cycle, but this is too frequent an effect to alter climate much. On longer time scales, there are no direct observations of the fluctuation of solar output. Observations and proxies for solar activity going back centuries or more do indicate long-term fluctuations in activity, as measured by sunspot number or solar-wind effects; it is not known how much fluctuation in solar brightness goes with such fluctuations in activity. It has been suggested that the solar influence on climate can be mediated directly by the influence of cosmic-ray fluxes (modulated by the solar wind's effect on the earth's magnetic field) on cloud formation (Svensmark and FriisChristensen, 1997; Svensmark, 1998), but the magnitude or even sign of the effect has not yet been quantified. Solar influences have been suggested as the cause of the Little Ice Age (Broecker, 1999), and indeed changes in solar activity do seem to line up with some major climate fluctuations. Solar forcing has been tied to drought frequency and effects on Mayan civilization (Hodell et al., 2001).

Exotica and Surprises

In addition to the well-studied mechanisms detailed above, there might be other threshold phenomena in the climate system that are difficult to assess quantitatively. The possibility of catastrophic release of methane by breakdown of frozen gas-ice compounds (clathrates) in permafrost or the ocean floor is in this category. Methane release has been clearly implicated in the warm event at the end of the Paleocene (Kennett and Stott, 1991; Dickens et al., 1995) (Box 4.1), and it has been argued that some of the Pleistocene methane signal is due to clathrate decomposition (Kennett et al., 2000) rather than tropical or high-latitude hydrological circumstances.

Furthermore, it must be acknowledged that the earth's climate system has in the more distant past exhibited major switches in mode of operation that are simply not understood. Notably, the past climate has oscillated between hothouse climates lasting tens of millions of years, when there was little or no permanent polar ice, and icehouse climates like those of the present and the Pleistocene. Both states have occurred throughout geological history. The most recent period of increased warmth continued throughout the Cretaceous (65 million years ago) into the Eocene (55 million years ago) and terminated with the onset of major ice ages about 2 million years ago. However, there were also icehouse periods earlier in the earth's history, including times during the Carboniferous and the Neo-Proterozoic. Although it is generally believed that geochemically mediated changes in atmospheric carbon dioxide played a major role in such transitions, there has been little success in reproducing the key features of hothouse climates by increased carbon dioxide alone. Concentrations of carbon dioxide high enough to prevent permanent polar ice in models generally lead to simulation of tropical oceans warmer than suggested by available data (Manabe and Bryan, 1985); the realism of both the tropical temperatures and the very high carbon dioxide levels are still under debate (Pearson et al., 2001). It had been hoped that better understanding of dynamic ocean heat transport would solve the problem, but recent work on Cretaceous and Eocene ocean dynamics does not support this idea. Moreover, even in simulations with increased carbon dioxide, continental interiors become too cold in the winters to reconcile with the equable climate that the fossil record demands. The problem of hothouse-icehouse transitions underscores that as-yet-unidentified mechanisms for mediating radical changes, some of which could well be abrupt, are lurking in the climate system.

Compounding the mystery of initiation and maintenance of the above "hot" mode of the climate is the growing evidence that the earth has fallen into an extremely cold "snowball-earth" state, in which the entire planet became ice-covered. The most recent occurrence of a snowball state was in the Neo-Proterozoic, about 600 million years ago. The circumstances in which the Snowball can be triggered are hotly debated but almost certainly involve reduced solar intensity, low carbon dioxide, ocean heat transport, and dynamics of sea ice (Hoffman et al., 1998; Poulsen et al., 2001; Hyde et al., 2000).

ABRUPT CLIMATE CHANGE AND THERMOHALINE CIRCULATION

The previous section explored a variety of mechanisms that might be involved in abrupt climate change. However, because sudden change in the THC stands as the only well-developed theory to explain abrupt climate changes, such as the Younger Dryas and the Dansgaard/Oeschger and Heinrich events, we now investigate this phenomenon in greater detail. This is not intended to imply that other mechanisms in this rapidly evolving field will not be found that could also contribute substantially to abrupt climate change, but only that models for other such mechanisms are not as mature as for THC changes.

Processes Driving the Thermohaline Circulation

The global THC consists of: cooling-induced deep convection, brine rejection, and sinking at high latitudes; upwelling at lower latitudes; and the horizontal currents feeding the vertical flows. Contrary to widespread perception, convection and sinking are neither the same nor co-located (Marotzke and Stott, 1999) because when rotational effects are strong, flow tends to be around a patch of maximal surface density (characterizing convection) rather than into it (Marshall and Schott, 1999). In the North Atlantic, where much of the deep sinking occurs (Gordon, 1986; Ganachaud and Wunsch, 2000), the THC is responsible for the unusually strong northward heat transport; part of this heat is imported from the Southern Hemisphere. Much of this heat is given off to the atmosphere over the Gulf Stream, from where it is transported northeastward by the atmosphere. This part of the heat loss is typical of all subtropical gyres and is not associated with the global overturn (Talley, 1999). The enhanced heat transport has been believed by many to contribute to the relative mildness of western European climate, particularly that of Scandinavia. However, as described earlier in this chapter, the relative contributions to European climate of the THC, the wind-driven ocean circulation, atmospheric transport associated with land/ocean contrasts, atmospheric planetary waves, and so on, remain uncertain.

The THC is maintained by density contrasts in the ocean, which themselves are created by atmospheric forcing (air-sea heat and water fluxes) and modified by the surface circulation. A crucial question is which density contrast one should consider—the one between the equator and the poles, or the one between North and South Atlantic, or perhaps even between North Atlantic and North Pacific. The choice matters in assessing what order of magnitude of change in surface density it might take to change the THC drastically. "Pole-to-pole" density differences are about 1 order of magnitude smaller than "pole-to-equator" ones and hence much more easily influenced.

Surface-density contrasts can be influenced by a wide variety of processes, both internal to the ocean and coupled to the atmosphere. For example, the THC transports relatively warm and salty waters from the subtropical North Atlantic into the convection regions, with opposite effects on density. The import of warm water tends to reduce the density in the high latitudes, and the import of saline water increases density. An indirect effect is associated with evaporation (and later export of water vapor), which occurs preferentially over warm water. Analysis of atmospheric data suggests that the Atlantic drainage basin loses moisture to the Pacific (Warren, 1983; Zaucker and Broecker, 1992). The resulting accumulation of salinity in the Atlantic is compensated for by a net influx of freshwater from the Southern Ocean to the Atlantic. Changes in the water balance of the tropics (for example, changing El Niño patterns) might influence the THC if sustained long enough (Schmittner et al., 2000; Latif et al., 2000). Another important driver is sea ice (Aagard and Carmack, 1989), which is important for the freshwater budget of the North Atlantic convection regions. Thus, the central question in understanding and simulating the role of THC changes in abrupt climate change is: What is the combined effect of all these feedback mechanisms in a climate-change scenario?

An understanding of abrupt climate change thus requires a detailed quantitative knowledge about the various driving processes and their combined effect on the THC. In principle, a change in any of these processes can generate substantial climate change in areas influenced by the THC. However, those climate changes will not necessarily be limited to those areas. Effects could be widespread, and vary from region to region, because of changes in patterns of natural climate variability (such as NAO and ENSO) and their associated teleconnections. This is largely unexplored terrain that needs enhanced research.

A Model Hierarchy to Investigate Abrupt Climate Change

Research by Bryan (1986) marked the beginning of realistic process modeling of the THC. This work showed that the THC in a three-dimensional ocean model can assume multiple equilibria with vastly different locations of deep-water formation. Marotzke and Willebrand (1991) found four qualitatively different equilibrium solutions in an idealized global model of the THC. Common to these models was a special formulation of the surface boundary conditions that took into account the different response characteristics of the sea surface to changes in atmosphere-ocean heat and freshwater fluxes. Multiple equilibrium solutions were reported also in a fully coupled atmosphere-ocean model (Manabe and Stouffer, 1988); thus, the result does not depend on the specific simplifications used in the ocean-only models.

Slow changes of the surface freshwater balance constitute one possible mechanism to induce abrupt change. Using an ocean-only model, Mikolajewicz and Maier-Reimer (1994) demonstrated that switches in the THC occurred if the discharge of freshwater to the Atlantic exceeded a threshold value. Other coupled models used freshwater pulses to disturb the circulation and exhibit responses that range from a large reduction (Manabe and Stouffer, 1997) to a full shutdown of the THC (Mikolajewicz et al., 1997; Schiller et al., 1997).

These models suggest that the large changes observed in the paleoclimatic records were due to rapid changes in the THC. A more systematic investigation, however, was difficult to perform with the models because of their high computational burden. Extensive parameter studies, the basis of the advancement of understanding, are hardly possible. In recent years, simplified climate models—or climate models of reduced complexity—have been developed (Stocker and Marchal, 2001). They contain limited dynamics and have high computational efficiency. This was a crucial step forward in extending the toolbox to investigate abrupt climate change. Such models are also referred to as "earth-system models of intermediate complexity." There are three strategies for formulating such models:

1. Rigorous reduction of the governing equations of the climate system.
2. Combination of model components of differing complexity.
3. Mathematical linearization of the response of comprehensive climate models to pulse perturbations.

Only strategies 1 and 2 are applicable to the problem of abrupt climate change, because of the nonlinear nature of the phenomenon. Coupled climate models of reduced complexity have been obtained with either strategy. Following the first, by zonally averaging the equations of motion in the ocean (e.g., Marotzke et al., 1988; Wright and Stocker, 1991), a very efficient ocean model component is obtained, which can be coupled to an energy balance model of the atmosphere (Stocker et al., 1992) or a statistical-dynamical model of the atmosphere (Petoukhov et al., 2000). The choice implies that the focus of investigation is restricted to the latitude-depth structure of the flow and to *a priori* chosen time scales that are accessible with these models. The second strategy was followed when three-dimensional ocean-circulation models were combined with a latitude-longitude energy balance model (Fanning and Weaver, 1997) or with an atmospheric-circulation model of reduced complexity (Opsteegh et al., 1998). Such combinations can be integrated for many thousands of years, thanks to the relative simplicity of the atmosphere.

Overall, models of reduced complexity are highly useful tools in paleoclimate research, and in particular for investigations of abrupt climate change, provided that they are used wisely. Clearly, they cannot replace general circulation models (GCM), because the reduced-complexity models consider only a limited set of constraints that are important in the climate system. The weaknesses of the reduced-complexity models are the incompleteness of dynamics and their often reduced resolution. Their strength is their computational efficiency, which permits extensive sensitivity studies or ensemble or even Monte Carlo simulations. Single simulations with reduced-complexity models are not useful to advance the science even if they happen to agree well with paleoclimatic data. However, these models are key tools in the process of quantitative hypothesis-building and -testing, not only in paleoclimatology but also in climate dynamics in general, because they allow the investigation of some feedback or process in its purest, isolated form. Often, the understanding gained from the reduced model is used to interpret the results from complex models. In addition, results from simple models or conceptual considerations have helped to define the strategy pursued in the use of complex models. In the following section, the committee adopts this approach to investigate the most fundamental questions concerning the stability of the THC and its role in abrupt climate change.

Abrupt Change, Thresholds, and Hysteresis

According to the definition given in Chapter 1, a climatic response faster than a change in forcing would be called abrupt. Such a rapid response would not occur as a result of small perturbations about a reference state, but only if a threshold was crossed; after that, a new state would be rapidly approached. Systems that exhibit such behavior often show hysteresis; that is, even if the perturbation has ceased after leading to the crossing of a threshold, the system does not return to its original state. Box 1.1 showed a

Box 3.1
Thresholds and Hysteresis

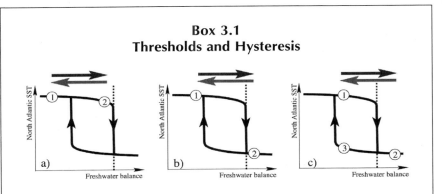

FIGURE 3.1

Many simple physical systems exhibit abrupt change, as demonstrated in the simple diagram presented in Box 1.1. The more complex figure (Figure 3.1) provides a schematic view of hysteresis in the thermohaline circulation. The upper branch denotes climate states in which the THC is strong and North Atlantic temperatures are relatively high (similar to present conditions). The lower branch represents a much-reduced or collapsed THC, in which the Atlantic meridional heat flux by the ocean is small. A given perturbation (indicated by the horizontal arrows) in the freshwater balance of the North Atlantic (precipitation plus runoff minus evaporation) first causes transitions from an initial state 1 to state 2. The reverse perturbation then causes a transition back to state 1, or to state 3. Three structurally different responses are possible for the same pair of perturbations, depending on whether threshold values (dashed line) are crossed. This, in turn, depends on where state 1 is, relative to the threshold: a) small, reversible response; b) large, reversible response; c) large, irreversible response (Stocker and Marchal, 2000).

mechanical analogue of such behavior; Box 3.1 presents a sketch of hysteresis of the THC resulting from perturbations in the freshwater forcing.

The simplest variant of a hysteresis loop is a useful tool to discuss the different possibilities for how the climate system can respond to changes in some controlling variable. More freshwater increases the buoyancy of the surface waters and tends to reduce the strength of the THC; less freshwater makes the THC more stable. This is illustrated by the change in the location of the system on the upper branch of the hysteresis. As long as perturbations do not exceed thresholds, the system response is weak (Figure 3.1a). An abrupt change is triggered on the crossing of a threshold. A small change in forcing can then cause large additional perturbations (Figure 3.1b). If the initial state of the ocean-atmosphere system is a unique equilibrium, the system jumps back to the original state once the perturbation has ceased; the abrupt change is reversible. However, if other equilibria exist, the perturbation can cause an irreversible change (Figure 3.1c), unless a perturbation is applied that has the opposite sense of the original one and is large enough; in Figure 3.1c, state 3 would have to be pushed to the left, beyond the upward-pointing branch.

Experiments with simplified ocean-circulation and climate models have helped to discover the possible hysteresis behavior of the atmosphere-ocean system. As shown in Box 3.1, hysteresis is one manifestation of multiple equilibria in a nonlinear system. The existence of hysteresis for the THC was first shown by Stocker and Wright (1991), who used a simplified model. For some values of the freshwater balance of the North Atlantic, the THC can be either in a strong or in a collapsed state. Numerous studies with a variety of ocean models coupled to simple representations of the atmosphere have demonstrated the existence of hysteresis (e.g., Mikolajewicz and Maier-Reimer, 1994; Rahmstorf and Willebrand, 1995); this is a robust property of such models. Obviously, in more-complex models, the hysteresis can consist of a number of sub-branches nested in a complicated way. However, it is unclear whether the hysteresis behavior would persist in more-realistic coupled models, particularly if the ocean component has spatial resolution believed to be necessary to be quantitatively consistent with observations. Likewise, it is unclear where the climate system is now on the hysteresis curve of the Atlantic THC: What is its structure? Does it have thresholds? If so, how close is the threshold? The following discussion demonstrates how model- and parameter-dependent the answer can be.

Abrupt Climate Change in a Minimal Model of the Thermohaline Circulation

The response of the THC to a perturbation of the surface freshwater balance can be illustrated in the two-box model of Stommel (1961), which is explained in Box 3.2. A new addition to the original system is horizontal diffusion. It reflects the transport of salinity by the ocean gyres, the mid-latitude systems of ocean circulation characterized by swift currents near the western boundary (for example, the Gulf Stream) and slower return flows farther eastward, occurring at virtually the same depth. Diffusion has a dramatic influence on the structure of the solutions (Box 3.3). For weak diffusion, the model exhibits hysteresis, which enables the existence of abrupt change in response to slow changes in the forcing; strong diffusion eliminates hysteresis (Plate 6). As shown in Plate 6, from their arbitrary starting point, both models initially migrate toward the steady state for the freshwater forcing H = 0.2, where large H indicates large transfer of freshwater through the atmosphere from the low-latitude to the high-latitude ocean (or its computational equivalent, large transfer of salt through the atmosphere from the high-latitude to the low-latitude ocean). When H is increased slowly, both models follow their equilibrium curves, as indicated by the arrows. The standard model has a threshold at H = 0.3, and an abrupt transition towards the other, "reverse" equilibrium occurs (orange curve). From then on, the model system follows the lower equilibrium curve; the hysteresis is shown by the orange curve remaining on the lower, red branch, even after the freshwater forcing has returned to its original value of 0.2. It would require a reduction of H to below 0.1 to force a return to the upper branch; if H stays above 0.1, changes remain permanent even after the perturbation of H is removed. The response of the diffusive model to changes in H is completely different (green curve). At each instant, the change in the THC scales with the forcing and no abrupt transition is observed. This model version has only one equilibrium solution for any given freshwater flux H, and, in contrast with the version with multiple equilibria, it exhibits only reversible changes (Plate 6).

A different way of illustrating this behavior is depicted in Figure 3.4, which shows the time evolution of the THC in response to a slow increase in freshwater forcing followed by an equally slow decrease. The diffusive model approaches zero THC strength essentially on the time scale of the change in forcing. In contrast, the standard case starts out with a slow

Box 3.2
A Minimal Model for the Thermohaline Circulation

For many years, the textbook example for explaining multiple states of the THC has been an idealized two-box model confined to a single hemisphere (Stommel, 1961), although a two-hemisphere version (Rooth, 1982) would be more appropriate to describe the Atlantic THC (Marotzke, 2000). Nevertheless, Stommel's configuration (Figure 3.2) is useful to classify existing simulations of abrupt climate change involving the THC and to identify some crucial open questions concerning the future stability of the Atlantic's THC.

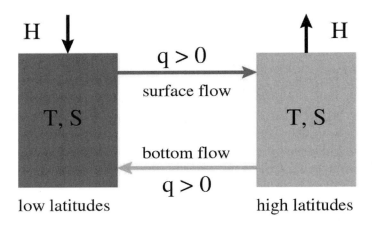

FIGURE 3.2

decrease as well but then responds with a sudden transition to the reverse mode. This is an abrupt climate change according to the definition used in this document; the response is considerably faster than the change in forcing.

As for the real Atlantic and climate system, the crucial question is whether the THC is best characterized by the "lower" or by the "upper" case. Is there a possibility that a temporary perturbation in, say, the freshwater forcing can induce a permanent change in the THC? This would

Figure 3.2 shows the original set-up of Stommel's two-box model for the THC. North-south exchange of water between a low-latitude box (left) and a high latitude box (right) is parameterized as a function of the density difference between the two boxes. H denotes the freshwater forcing, the flux of freshwater through the atmosphere from low latitudes to high latitudes or equivalently, as shown here, the flux of salt through the atmosphere from high latitudes to low latitudes.

The circulation consists of a volume flux q taken to be proportional to the density difference between high and low latitudes. If q > 0, there is poleward surface flow because high-latitude density is greater than low-latitude density, and vice versa. At low latitudes, the ocean gains heat from and loses freshwater to the atmosphere; the opposite is true at high latitudes. Consequently, both temperature and salinity are higher at low latitudes than at high latitudes. This has opposite effects on density. When q > 0, the temperature difference dominates the density difference and drives the circulation, whereas the salinity difference brakes it, and vice versa. The situation q > 0, with sinking implied at high latitudes, is familiar from the North Atlantic and describes today's active THC.

In a plausible limiting case (Marotzke, 1990), the box temperatures are assumed to be imposed by the atmosphere, as is the surface freshwater exchange. A new addition to the system is horizontal diffusion. It reflects the transport of salinity by the ocean gyres, the systems of ocean circulation characterized by swift currents near the western boundary and slower return flows farther eastward, occurring at virtually the same depth. Understanding the effect of ocean gyres on the stability of the THC is crucial, and the models currently used are likely to distort this influence because of their low spatial resolution.

Two cases are considered here. The "standard" case is the classical THC box model with only very weak diffusion; the other case will be designated "diffusive." In the example shown, the diffusive case has a different proportionality factor relating density differences to flow strength, such that with the same freshwater forcing, the two cases have very similar strengths of the "normal" North Atlantic THC.

imply that the real THC displays hysteresis and the possibility of abrupt change. Or would any transition be smooth, on the time scale of the forcing change, and be reversible? Simulations with comprehensive three-dimensional GCMs show both types of behavior. An ocean model with relatively large diffusion, caused by the choice of the numerical scheme, exhibits a transient behavior very similar to the upper dashed curve in Figure 3.4 when a slow freshwater flux perturbation is applied to the North Atlantic (Mikolajewicz and Maier-Reimer, 1994). A less-diffusive coupled model

Box 3.3
Multiple Equilibria and Model Parameters

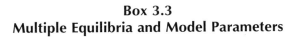

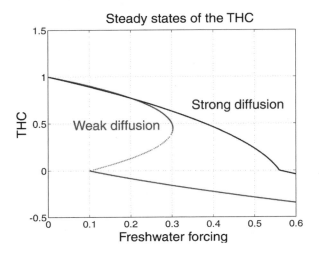

FIGURE 3.3   Steady-state solutions of the two-box model as a function of the atmo-sphere-ocean freshwater flux H. Depending on the strength of the horizontal mixing (diffusion) the model exhibits multiple eqilibria for a limited range of freshwater fluxes (bottom curve) or, alternatively, a unique solution (top curve). (See plate 6 for more detail.)

exhibits a permanent change in the THC (Manabe and Stouffer, 1988). This aspect has been associated with the quantitatively different vertical mixing in these models (Manabe and Stouffer, 1999).

The simple box model above demonstrates that the shape and position of the hysteresis in parameter space depend strongly on the model setup and on values of model parameters. Uncertainties in the parameters translate directly into uncertainties in the hysteresis and therefore uncertainties in the likely response of a system to a perturbation. Investigations using simplified models indicate that mixing schemes and values of mixing parameters are crucial in determining the shape of the hysteresis (Knutti and Stocker, 2000; Schmittner and Weaver, 2001). Unfortunately, mixing in the ocean is not well simulated in current-generation climate models, because of poorly parameterized small-scale processes and insufficient resolution. The location of the present, past, and future states of the climate system on the hysteresis curve, and its shape

Figure 3.3 (also see Plate 6) shows that the steady states of the two-box model can be calculated analytically and the dependence of the flux q (a measure of the strength of the THC) on the freshwater flux H can be examined. In the standard case (weak diffusion, lower curve), the THC is strongest (value arbitrarily set to 1) when the freshwater flux H vanishes. With higher H, the THC weakens. If H = 0.3, no steady-state solution with q > 0 is possible. However, for H > 0.1, there is a second stable equilibrium, a reverse mode of the THC, with q < 0. This flow pattern strengthens as H increases. For a certain parameter range, here 0.1 < H < 0.3, three equilibria are possible; it is readily shown that the middle one (on the dotted part of the curve) is not a stable solution. In that range of H, the model exhibits hysteresis.

The presence of hysteresis is strongly dependent on model parameters. This is shown for a case in which the effect of the horizontal mixing due to gyre transports is increased (strong diffusion). Hysteresis disappears (upper curve); and for progressively increasing freshwater forcing, the THC smoothly approaches zero and—again smoothly—turns into the reverse mode for H = 0.56. For any given freshwater forcing, there is a unique and stable THC. The presence of multiple equilibria of the THC therefore depends strongly on the model formulation, parameterization of processes, and choice of parameter values.

and complexity, clearly are among the major unsolved problems in climate dynamics. Results that depend crucially on a specific shape of the hysteresis are likely not to be robust findings at this stage.

Whether the nonlinearities giving rise to abrupt change in the simplified models are an artifact of the simplifications or carry over to more complex and realistic systems needs to be investigated. Only more comprehensive and more complete climate models can make a convincing case that the assumptions underlying the nonlinearities in these simple models bear sufficient realism.

Simulation of Past Changes of the Thermohaline Circulation

Earlier, three fundamental ways of causing abrupt climate change were presented; all have been simulated with coupled GCMs. The first category

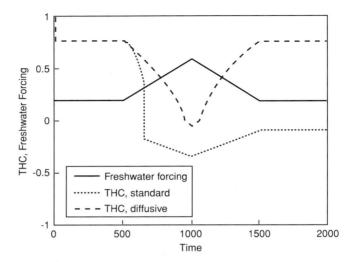

FIGURE 3.4   Response to the thermohaline circulation (THC) to a slow increase and subsequent slow decrease in freshwater forcing for the two cases in Figure 3.3.

concerns the response to a sudden external perturbation, such as a disintegrating ice sheet. Manabe and Stouffer (1995) released a short, strong pulse of freshwater (1 Sv = 10^6 m³/s for 10 years) into the northern North Atlantic in their coupled model. The THC responded instantaneously with a reduction of almost 70 percent within a decade but recovered to the original strength within less than 200 years. Although the THC shut down almost completely, a threshold was not crossed, and no transition to a new equilibrium was simulated. The model does, however, have a second equilibrium that is achieved if the perturbation is much stronger (Manabe and Stouffer, 1999). It appears, therefore, that the behavior of the model is qualitatively similar to the standard case of the Stommel two-box model (see Plate 6, orange line). Depending on the amplitude of the perturbation, the coupled GCM thus exhibits either a transition to a new stable state or a reversible change (which might still be abrupt and of large amplitude). Manabe and Stouffer (1999) also showed that simply increasing the vertical diffusivity in their model led to a situation with a single equilibrium; this implies that processes other than those discussed above could qualitatively change the hysteresis structure of the model.

Behavior similar to that in Manabe and Stouffer's (1995) model was observed by Schiller et al. (1997), albeit because of a different forcing. They forced their model with a freshwater flux qualitatively similar to that in Figure 3.4. This coupled model appears to have a unique equilibrium, and the simulated changes evolve roughly on the time scale of the forcing. Any abruptness in such a model will therefore be generated only through the forcing and not through the dynamics of the model itself. The transient behavior of this complex model is qualitatively comparable with that of the diffusive Stommel model (Figure 3.4, upper dashed curve). The complete, temporary shutdown of the THC causes a massive cooling in the North Atlantic and weak warming in the South Atlantic. The climatic behavior of opposite sign during a full THC shutdown has been termed the bipolar seesaw (Broecker, 1998; Stocker, 1998), and is a phenomenon that was probably active during most of the last glacial period (Blunier and Brook, 2001) (see Plate 2).

The second type of abrupt change in climate models occurs when slow changes in the forcing push the model beyond a threshold and induce a transition to a second equilibrium. This behavior is simulated in ocean-only models that include simplified formulations of the atmosphere (Rahmstorf, 1994, 1995) and in models of reduced complexity (Aeberhardt et al., 2000; Schmittner and Weaver, 2001). Periodic forcing can also trigger abrupt change in a model that has multiple equilibria, provided that the forcing amplitude covers both branches of the hysteresis. Ganopolski and Rahmstorf (2001) showed that weak forcing is sufficient if the model's hysteresis has a narrow shape. As emphasized earlier, such results are not considered robust, because they depend strongly on the structure of the hysteresis and the location of the initial state on it. Tuning and choices of parameters can strongly influence the transient behavior under a given forcing (Schmittner and Weaver, 2001) (Figure 3.4).

The third possibility for the occurrence of abrupt climate change in model simulations is a spontaneous regime transition similar to the paradigm of the Lorenz (1963) model. This concept has been explored in simplified models (Timmermann and Lohmann, 2000) but was recently also found in a control run of a coupled atmosphere-ocean GCM (Hall and Stouffer, 2001). Annual mean surface air temperature, averaged over a region extending from southwestern Greenland to Iceland, showed natural interannual variability of around ±1°C (Figure 3.5, top). During the 15,000-year integration, one event occurred in which the temperature underwent a rapid decrease of about 4°C within a few decades (Figure 3.5, bottom). The event was triggered by an atmospheric pattern of persistent northwesterly flow in the region. This induced a southwestward Ekman flow in the surface ocean,

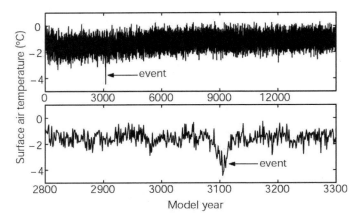

FIGURE 3.5   Model of surface air temperature averaged over an area spanning from southwestern Greenland to Iceland found in a 15,000-year control integration of a coupled atmosphere-ocean Global Circulation Model. The top shows the entire series, while the bottom highlights a 500-year period that contains an abrupt cooling event. (Hall and Stouffer, 2001).

enhanced the East Greenland Current, and brought freshwater into the northern North Atlantic. Only then did the THC respond with a weakening that amplified the initial perturbation. In this example, changes in the THC were a response to rather than a cause of the cooling event. Although the result is intriguing, its robustness is poorly understood; the lack of intermediate-size events during the long integration is a cause for concern.

Some Open Questions

Even if one accepts that the THC played a central role in past abrupt climate change, the above examples of recent studies underline how our understanding of past rapid changes is still limited. In particular, the question of what triggered abrupt climate change, such as the Younger Dryas, has not yet been answered, unless we assume that the climate system essentially exhibits spontaneous regime transitions of the THC. In the following, we attempt to break down that overarching question into smaller, more manageable topics. The problem contains some fundamental aspects, as epitomized by the two different hysteresis structures in Plate 6, indicating that the whole spectrum of approaches—theory, modeling, and observations—will be needed.

The conceptual discussion above has been framed entirely in terms of Stommel's hemispheric model of the THC, but the likely importance of the bipolar see-saw alone indicates that this view must be broadened. Two seemingly unrelated questions help to focus the discussion: First, what is the role of convective mixing in the THC? Second, if salinity is so important in determining the global THC pattern, why does the North Atlantic gain its high-latitude surface density mainly through heat loss (Schmitt et al., 1989)? The first question is motivated by the strong observational (Munk, 1966; Munk and Wunsch, 1998) and numerical (e.g., Bryan, 1987; Marotzke, 1997) evidence that the strength of the THC is strongly controlled by the vigor of vertical mixing in the stratified regions of the ocean. This is because the deepwater that upwells in these regions must be heated by mixing to maintain the near-surface higher temperatures. In contrast, the THC strength is quite insensitive to the efficiency of convective mixing (that is, the speed with which the ocean eliminates dense water overlying lighter water), and the THC could even be strong in the absence of convective mixing, according to the single-hemisphere GCM of Marotzke and Stott (1999).

The answer to the first question leads the way to answering the second. Marotzke (2000) has argued that the results from single-hemisphere models should be viewed as applying to the global integral of the various THC branches and that the magnitude of the global integral obeys different laws from the distribution of the grand total over various competing deepwater formation sites. A number of idealized and more realistic ocean GCMs have shown that varying the freshwater flux forcing leaves the global integral of deep sinking nearly constant but the strength of North Atlantic sinking considerably changed (Tziperman, 1997; Klinger and Marotzke, 1999; Wang et al., 1999). Hence, a *reduction* in Northern Hemisphere THC would be associated with an *increase* in Southern Hemisphere THC. Thus, the globally integrated THC (sum of all branches) is rate-limited by vertical mixing and the gross pole-equator density contrast, which is dominated by the pole-equator temperature contrast (not by salinity). It follows that convection is basically driven by heat loss and that it occurs predominantly at high latitudes.

That so much oceanic deepwater is formed in the North Atlantic, and not in any of the other competing high latitude regions, is determined by the North Atlantic's high surface salinity and hence high surface density. These, in turn, are strongly influenced by the freshwater flux forcing; moreover, the deepwater formation is intimately related to the very deep convective mixing occurring there.

This two-step procedure—considering first the global integral of THC strength and then its distribution over different regions—helps to sort out a number of conceptual questions, but others arise. Most important is the role of the Antarctic Circumpolar Current, especially the wind-induced upwelling there (Toggweiler and Samuels, 1995); this has only recently begun to be addressed theoretically (Gnanadesikan, 1999). The interaction of the North Atlantic THC with the other oceans is poorly understood, both conceptually and from observations (e.g., Whitworth et al., 1999).

Extending one's perspective beyond the Stommel model also calls into question the long-held tenet that freshwater forcing necessarily weakens the THC. Rooth's (1982) interhemispheric box model suggests that the Atlantic THC actually *increases* with increased freshwater flux (Rahmstorf, 1996; Scott et al., 1999; Marotzke, 2000). This is confirmed for the Atlantic branch of the THC in an idealized global GCM, as long as one considers the equilibrium response (Wang et al., 1999). Under a faster increase, as would be expected with increased greenhouse gases or might have occurred with outburst floods or ice-sheet surges in the past, the THC did weaken.

If the equilibrium response can be interpreted as reflecting the THC's response to very slowly varying atmospheric moisture flux, as might plausibly have happened during the glaciations, the assumed overall glacial weakening of the THC, as opposed to the shorter-lived and stronger proposed weakening of the THC in specific events during the glacial, could be explained.

Given confirmation that the North Atlantic surface densities and the resulting convective activity do matter, the question arises whether we understand, and can observe, what they are influenced by. The drivers could be oceanic transports of freshwater (e.g., Aagard and Carmack, 1989) and heat, local surface fluxes, or remote influences, such as the water-vapor transport from the Atlantic to the Pacific (Zaucker and Broecker, 1992; Schmittner et al., 2000; Latif et al., 2000). In addition, one should consider the evolution of surface density in the Southern Ocean, in particular the Weddell Sea, because it is not well observed and the processes that link it with North Atlantic densities are not well understood.

There is a huge gap in our conceptual understanding linking changes in convective activity, in the North Atlantic or elsewhere, to the THC and the northward heat transport. Regionally, changes in properties appear to occur rapidly but are poorly understood (e.g., Sy et al., 1997). On a larger scale, the THC (and hence heat transport) can respond to changes in external forcing much faster than on the millennial time scale of thermodynamic equilibration of the deep ocean, but it is not clear which of the various plausible, identifiable time scales—ranging from months to decades—is most relevant. The classical picture (Kawase, 1987; Döscher et al., 1994) suggests that the deep circulation is set up through wave processes on time scales of months to a few years, but it has also been argued that advection by the deep western boundary current must be important, and this implies time scales of decades (Marotzke and Klinger, 2000).

The potential predictability and prediction of the THC raise the thorny issue of assessing the quality of numerical simulations of future climate evolution. The fundamental problem is best understood when juxtaposed with daily weather forecasting. Through thousands of forecasts based on model simulations, it has been established that weather predictions have "skill"; that is, they improve upon a naïve baseline. But the time needed to test a weather forecast typically is a day—we will know tomorrow night whether tomorrow's picnic gets rained out. An analogous accumulation of evidence of forecast skill is impossible when the prediction lead time is decades and more. How, then, can we state with well-defined confidence what is likely in store?

An abrupt THC change is likely to have consequences massive enough that its probability should be estimated, if only crudely. In lieu of establishing true forecast skill, the models used for future climate-change scenarios should pass all the tests posed by available data. For example, one would reasonably have more confidence in climate models of future evolution that were able also to simulate the complicated climate history, including the rapid changes, of the last glacial interval. However, it is difficult to establish the "transferability" of putative modeling success of the past. Although being able to simulate the Younger Dryas is neither strictly necessary nor sufficient for being able to predict the THC's evolution under global warming, it would enhance confidence considerably. Another strategy would be to simulate an ensemble of possible future climates and ascribe probabilities to them; an example is discussed in Chapter 4. Still, the problem remains of assessing whether the probabilities have the right order of magnitude. A further limit to predictability might arise from chaotic dynamics. In the global model of Wang et al. (1999), the timing and presumably the critical threshold of a THC collapse were fundamentally unpredictable to within a factor of 2. Different realizations of statistically identical random perturbations in the wind field in Wang et al. (1999), mimicking different (unpredictable) weather histories, led to collapse times of that varied between 250 and 500 years. These results have not been confirmed (or challenged) with more sophisticated models, and the random perturbations might have been unrealistically large, but the possibly chaotic nature of THC changes needs to be addressed.

Considering the known limitations of climate models, it is not currently possible to ascribe probabilities to future abrupt climate changes. Given the possible large impacts, simulation of abrupt climate change caused by a collapse in the THC constitutes a "Grand Challenge" problem in computational earth science. A dedicated supercomputer could be used to test whether the THC shows the possibility of threshold behavior, abrupt change, and hysteresis in a climate model that in its ocean components resolves the most important features of the ocean circulation. The conceptual models discussed here demonstrate that the outcome is open: on the one hand, high-resolution models are expected to have stronger gyre transport of heat and salinity, which appears to work in favor of the diffusive paradigm; on the other hand, a high-resolution model would be expected to exhibit a greater degree of internal variability, increasing the chance of self-induced threshold-crossing—if indeed a threshold exists. The need for substantially increased computational resources is clear.

4

Global Warming as a Possible Trigger for Abrupt Climate Change

he review of possible mechanisms in Chapter 3 showed that in a chaotic system, such as the earth's climate, an abrupt climate change always could occur. However, existence of a forcing greatly increases the number of possible mechanisms. Furthermore, the more rapid the forcing, the more likely it is that the resulting change will be abrupt on the time scale of human economies or global ecosystems. Although abrupt climate changes have shocked ecosystems and societies over the last few millennia, these climate changes have not been as dramatic as those that occurred during the last ice age. It is probably no coincidence that stability of the climate increased when ice-sheet size and atmospheric carbon dioxide concentration largely leveled off at the end of the ice age.

Greenhouse gases are accumulating in the earth's atmosphere and causing surface air temperatures and subsurface ocean temperatures to rise. It is now the consensus of the science community that the changes observed over the last several decades are most likely in significant part the result of human activities and that human-induced warming is expected to continue (NRC, 2001). As discussed in Chapters 2 and 3, the abrupt climate changes of the past were especially prominent when orbital processes were forcing the climate to change most rapidly during the cooling into and warming out of the ice age, consistent with the results from modeling that forcing of climate increases the possibility of crossing thresholds that trigger abrupt change (e.g., Boxes 3.1 and 4.1). Given our understanding of the climate

---

Box 4.1
Abrupt Climate Change in the Ancient Hot House

About 55 million years ago, the earth underwent an abrupt climate change known as the Late Paleocene Thermal Maximum (LPTM), recently renamed the Paleocene-Eocene Thermal Maximum (PETM) (Kennett and Stott, 1991; Dickens et al., 1995; Dickens, 1999; Norris and Röhl, 1999; Bains et al., 1999). Against the backdrop of an already warm climate with reduced pole-equator temperature contrasts, bottom water temperatures increased by 4-6°C (Thomas and Shackleton, 1996), and high-latitude surface temperatures by 4-8°C, over 10-20 kyr (Norris and Röhl, 1999). Thirty to fifty per cent of benthic foraminifera went extinct (Thomas and Shackleton, 1996). The suite of dramatic global changes inferred for the LPTM includes increased aridity in subtropical latitudes and increased high-latitude precipitation (Robert and Maillot, 1990; Schmitz et al., 2001).

At the onset of the LPTM warming, marine and terrestrial carbon isotope values exhibit a negative shift of at least 2.5 per mil (Norris and Röhl, 1999). The only known sources for this quantity and composition of carbon today are the vast reserves of natural gas hydrate in oceanic, deep lake and polar sediments and the free methane gas trapped beneath hydrate deposits. Methane hydrate is a solid complex of methane and water that is stable only at low temperatures and high pressures, such as found in sediment of the mid-depth and deep ocean. The carbon isotope signature during the LPTM is indicative of massive destabilization of marine methane hydrates (Dickens et al., 1995); it is estimated that 1,200-2,000 gigatons of methane gas were released.

---

system and of the mechanisms involved in abrupt climate change, this committee concludes that human activities could trigger abrupt climate change. Impacts cannot be predicted because current knowledge is limited, but might include changes in coupled modes of atmospheric-ocean behavior, the occurrence of droughts, and the vigor of thermohaline circulation (THC) in the North Atlantic. More research is needed to better understand the relationship between human influences on climate, especially global warming, and possible abrupt climate change.

CHANGES IN NORTH ATLANTIC
THERMOHALINE CIRCULATION

A question of great societal relevance is whether the North Atlantic THC will remain stable under the global warming expected for the next few

The cause of the LPTM warming is unknown (Dickens, 1999) but has been speculated to be increased volcanism (Thomas and Shackleton, 1996) or low-latitude deepwater formation (Bains et al., 1999). Recently, Bice and Marotzke (in press) presented results from an ocean general circulation model indicating that a sudden switch of deepwater formation from high southern to high northern latitudes could have led to mid-depth and deep-ocean warming of around 5°C. The switch was caused by a slow increase in the atmospheric water cycle, as expected under increasing temperatures (Manabe and Bryan, 1985; Manabe and Stouffer, 1993), and consistent with LPTM sedimentary evidence (Robert and Maillot, 1990; Schmitz et al., 2001). The mid-depth warming displayed by the model could destabilize large volumes of methane hydrate in the depth range of 1,000-2,000m over much of the world ocean. The THC switch seen in the model of Bice and Marotzke (2001) and the inferred subsequent methane release are an abrupt climate change, according to the definition given in Chapter 1. When the freshwater forcing is reduced to pre-LPTM values, deepwater is again formed in the Southern Hemisphere, with a hysteresis characterized by case (b) in Figure 3.1.

The climate during the LPTM may well be a valid past analogue of the greenhouse world expected for the next several centuries (Dickens, 1999), despite the different continental configuration. The results of Bice and Marotzke (2001) indicate more severe potential consequences of a drastic change in THC, as might occur in a future greenhouse world (Manabe and Stouffer, 1993), than previously assumed.

centuries. A possible shutdown of the THC would not induce a new glacial period, as press reports suggested; however, it clearly would involve massive changes both in the ocean (major circulation regimes, upwelling and sinking regions, distribution of seasonal sea ice, ecological systems, sea level) and in the atmosphere (land-sea temperature contrast, storm paths, hydrological cycle, extreme events). The most pronounced changes are expected in regions that are today most affected by the influence of the North Atlantic THC (e.g., Scandinavia and Greenland).

Current knowledge of the evolution of the THC is summarized in the Third Assessment Report of the Intergovernmental Panel on Climate Change (2001b). Several comprehensive coupled climate models were run with a scenario of increasing greenhouse gas forcing for the next 100 years. Most models show a reduction in the THC in response to the forcing (Plate 7). This is due to enhanced warming of the sea surface in the high latitudes and a stronger poleward atmospheric transport of moisture, leading to more precipitation in the North Atlantic region. Those two effects, in concert, lead to an increase in buoyancy of the North Atlantic surface waters, which reduces the THC. Although the relative strength of the two mechanisms is debated and uncertain (Dixon et al., 1999; Mikolajewicz and Voss, 2000), most climate models seem to show a general reduction in the Atlantic THC in response to global warming.

The exceptions to this behavior remind us of the inherent uncertainties present in the simulations. It is not clear whether all relevant feedback mechanisms are considered properly in the current generation of climate models and whether their strength is simulated realistically. A simulation by Latif et al. (2000) suggested that changes in the El Niño-Southern Oscillation (ENSO) frequency and amplitude might change the freshwater balance of the tropical Atlantic in such a way that increases in buoyancy in the high latitudes are compensated for by drier (and hence more saline) conditions in the tropics. Gent (2001) reported on a simulation in which evaporation from a warmer sea surface in the North Atlantic is not compensated for by enhanced precipitation, and this simulation results in a stabilization of the THC. While it is not currently possible to decide which simulations are more realistic—those of Plate 7 showing a THC decrease or those that do not—the two simulations by Latif et al. (2000) and Gent (2001) illustrate that the quantitatively correct simulation of heat and freshwater flux changes is essential for the projection of the evolution of the THC under global warming.

However, there are other uncertainties regarding the fate of the THC. Research indicates that the realized warming and the associated changes in the hydrological cycle constitute a threshold for the THC (Manabe and Stouffer, 1993). Also, the rate of warming appears to influence the stability of the circulation (Stocker and Schmittner, 1997; Ahmad et al., 1997), because the ocean heat uptake is limited by mixing; faster warming in the atmosphere produces stronger vertical density gradients in the ocean, which tend to reduce the sinking. Faster warming makes the THC less stable to perturbations. Furthermore, both theoretical arguments (Marotzke, 1996) and model simulations (Tziperman, 1997) suggest that the THC becomes less stable when it is weaker (i.e., once reduced, the THC is more susceptible to perturbations). In the extreme case very close to a threshold, the evolution of the THC loses predictability altogether (Knutti and Stocker, 2001) (Figure 4.1) as discussed in the next section. It is intriguing that recent measurements show that an important part of the North Atlantic

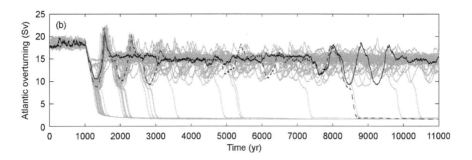

FIGURE 4.1 Evolution of the maximum overturning in the Atlantic (strength of the thermohaline circulation [THC]) for a coupled model of reduced complexity for 100 realizations of the model. Radiative forcing is increased from years 1000 to 1140, equivalent to a doubling of CO_2, and then held constant. The warming pushes the model closer to the bifurcation point, and transitions usually occur when the overturning is weakened. Two individual realizations are highlighted by the black lines (Knutti and Stocker, 2001).

THC (the Faroe Bank overflow from the Nordic Seas; Hansen et al., 2001b) has experienced both a reduction in total flow and a warming and freshening over the last 50 years, consistent with many model projections. However, the entire THC is not monitored. This highlights the importance of maintaining and increasing measurements and data-model comparisons that capture the behavior of the entire system.

LIMITED PREDICTABILITY CLOSE TO AN INSTABILITY

Possible instabilities of the THC also have important implications for the predictability of future climate change. Model simulations show that as an instability is approached, small deviations in initial or boundary conditions can determine whether a transition to a different equilibrium will occur, which inherently limits predictability. This behavior has been investigated with a climate model of reduced complexity (Knutti and Stocker, 2001). The threshold is approached by a prescribed global warming over about 140 years, equivalent to a doubling of carbon dioxide. Small random fluctuations, as produced by atmospheric disturbances at the ocean surface, can excite large changes in the THC when the system is close to a threshold (Figure 4.1). Many experiments with the same model but slightly different initial conditions (Monte Carlo simulations) indicate that the North Atlantic THC can undergo many oscillations before it settles in an active or a collapsed state. In some cases, a rapid collapse of the THC occurs many thousands of years after the perturbation. Obviously, beyond the problem of approaching an instability point and the increased vulnerability of the THC to further perturbation, such an evolution results in a much more unpredictable climate system (Figure 4.1).

These simulations are performed with a simplified model that includes only a limited set of processes. Experiments with comprehensive models are necessary to determine whether more realistic models exhibit similar behavior close to bifurcation points. In such models it would be crucial to identify whether there are regions or components in the climate system that could serve as early warning systems for a potential THC reduction or shutdown.

CHANGES IN NATURAL MODES OF
THE ATMOSPHERE-OCEAN SYSTEM

Data suggest that there has been an increase in the frequency of occurrence of El Niño conditions and a possible change in characteristic time scales of this phenomenon (e.g., Urban et al., 2000; Tudhope et al., 2001; see Chapter 2). However, the time series is too short to establish whether such rapid mode shifts are part of the natural, long-term variability inherent in this phenomenon or whether this is already a manifestation of a perturbed system. It has been suggested that this change in mode may be a response of the tropical Pacific to anthropogenic warming (Trenberth and Hoar, 1996).

Comprehensive climate models have just started to include credible representations of ENSO, which makes available tools to investigate the possible responses of the tropical Pacific atmosphere-ocean system to climate change. One model suggests that warming will cause El Niño events to become more frequent and stronger (Timmermann et al., 1999). Apart from regional implications, such a development could have far-reaching impact on other components of the climate system. For example, ENSO causes worldwide teleconnections, and it changes the freshwater balance of other ocean provinces (such as the tropical Atlantic). This provides a mechanism to modify the sea-surface salinity and hence the THC (Latif et al., 2000). Changes in the frequency and amplitudes of natural modes might not only evolve rapidly and thus manifest themselves as abrupt climate change, but they may also trigger other processes that lead to abrupt climate change.

As discussed in Chapter 2, strong trends have been occurring in the

North Atlantic Oscillation/Arctic Oscillation (NAO/AO). Among many possible causes, Shindell et al. (1999) suggested that these trends are a response to greenhouse-gas-induced warming. Large impacts could result from continuation and strengthening of such trends, perhaps leading toward a locking of the system in one of the preferred patterns (Palmer, 1999; Corti et al., 1999). Feedbacks on the THC have received the most attention (see below); other complex questions, including linkages with ENSO and other modes, merit greater attention in the future.

Most hypotheses with respect to possible abrupt changes in natural modes are based on but a few model experiments. Some of the models still have known biases (e.g., with respect to ENSO frequency). Nevertheless, such simulations indicate that changes in natural modes are a possibility when climatic conditions change.

Oceanic circulation modes may also experience major changes with greenhouse warming. Alteration of the THC thus can involve changes in structure in addition to a simple decrease or increase in amplitude. The present North Atlantic circulation involves two important modes of THC: the deep overflow branch originating in the Greenland-Iceland-Norwegian-Barents Seas, and the rapidly responding middle-depth THC driven by convection in the Labrador Sea (Plate 4). Wood et al. (1999), using a climate model with no flux corrections, suggested that greenhouse warming of the Nordic Seas between Greenland and Norway will cause reduction in density of the deep overflows. As they move south of Greenland they then will fail to circulate westward as boundary currents into the Labrador Sea. This, in turn, promotes a collapse of Labrador Sea deep convection, which is the primary driver of intermediate-depth THC (Häkkinen, 1999). The paleoclimatic record based on benthic and planktonic foraminifera (Hillaire-Marcel et al., 2001) suggests that Labrador Sea convection might indeed have been shut down for long periods during the previous interglacial warm period. This ironic scenario shows how the two modes of THC can interact, and in this case alter the balance in favor of the overflow mode.

POSSIBLE FUTURE ABRUPT CHANGES IN
THE HYDROLOGICAL CYCLE

One of the most striking predictions of climate models and theory is that global warming will put more moisture into the atmosphere in the tropics and generally accelerate freshwater transport to higher latitudes. Melting of land-fast ice, sea ice, and permafrost and biological or geological changes in drainage basins might provide extra abruptness to the freshwater cycle. Increasing precipitation and streamflow has been documented in North America. During the twentieth century, the zone 30°N to 85°N has experienced a 7-12 percent increase (Intergovernmental Panel on Climate Change, 2001a); this is the region critical to the Arctic and Atlantic Oceans. In 1998, the region north of 55°N was the wettest on record; and in the middle northern latitudes, precipitation has exceeded the 1961-1990 mean every year since 1995.

In global-warming scenarios, the GFDL coupled-climate model suffers a major decrease in THC amplitude owing to the resulting load of buoyant freshwater on top of the far northern Atlantic. The increase in precipitation minus evaporation (plus runoff from land) north of 45°N is nearly 50 percent under doubled carbon dioxide (Manabe and Stouffer, 1994). In other climate models, the freshening of the surface ocean is less strong, but high-latitude warming has the same effect. Adequate observations to establish crucial rates of the freshwater cycle are not being carried out.

One of the most important issues is the possibility of increase in extreme events related to land-surface hydrology. As summarized in Intergovernmental Panel on Climate Change (2001b), model projections of global warming find increased global precipitation, increased variability in precipitation, and summertime drying in many continental interiors, including "grain belt" regions. Such changes might produce more floods and more droughts. On the basis of the inference from the paleoclimatic record, it is possible that the projected changes will occur not through gradual evolution proportional to greenhouse-gas concentrations, but through abrupt and persistent regime shifts affecting subcontinental or larger regions. The inability to conduct long simulations with coupled models validated against paleoclimatic records, owing to resource limitations, leaves many uncertainties.

ICE SHEET CHANGES

Shrinkage or disappearance of all or part of a large ice sheet in response to natural or human-caused forcing remains a poorly quantified possibility. Changes in the balance between snowfall and melting at the ice-sheet surface might be important in the future (e.g., Intergovernmental Panel on Climate Change, 2001b), but attention is focused on changes in ice flow because they might affect sea level more rapidly and probably will be more difficult to predict.

The Heinrich layers in North Atlantic sediment indicate that rapid changes in flow of one of the ice-age ice sheets have delivered large quantities of icebergs to the ocean, perhaps over centuries and with substantial effect on sea level. Among the modern ice sheets, attention is focused, although not exclusively, on the West Antarctic ice sheet because its bed is in many places below sea level, well lubricated, and deeper toward the center of the ice sheet. These characteristics allow the possibility of West Antarctic flow instabilities (Weertman, 1974; Oppenheimer, 1998; Alley and Bindschadler, 2000). A sea-level change of about 5 m (from the West Antarctic ice sheet alone) (Plate 8) to perhaps more than 10 m (including response of the East Antarctic ice sheet to loss of the West Antarctic ice) in a few centuries may be possible, even if unlikely (Intergovernmental Panel on Climate Change, 2001b).

No prediction of change is yet available, and large uncertainties are attached to possible rates and magnitudes of change. Sedimentary evidence from beneath the West Antarctic ice sheet indicates that the ice sheet shrank substantially or disappeared at least once after it formed, although at an unknown rate (Scherer et al., 1998). The current, locally rapid changes in the West Antarctic ice sheet are difficult to explain, but the average change across the ice sheet is small (Alley and Bindschadler, 2000). Modeled behavior includes the possibility of large and rapid ice-sheet changes (MacAyeal, 1992), but behavior in other models is more stable (Hulbe and Payne, 2001). The major southern sites of formation of oceanic deep-waters are close to the West Antarctic ice sheet, and deep-water formation involves interaction with floating extensions of the ice sheet called ice shelves (e.g., Schlosser et al., 1994). The freshwater delivery of Heinrich events to the North Atlantic was associated with greatly reduced formation of deep-water in the vicinity and very large climate anomalies well beyond the region of freshwater supply (Broecker, 1994). Thus, any large changes in the ice sheets could affect many aspects of climate in addition to sea level (Plate 8).

OUTLOOK

If the increase in atmospheric greenhouse gas concentration leads to a collapse of the Atlantic THC, the result will not be global cooling. However, there might be regional cooling over and around the North Atlantic, relative to a hypothetical global-warming scenario with unchanged THC. By itself, this reduced warming might not be detrimental. However, we cannot rule out the possibility of net cooling over the North Atlantic if the THC decrease is very fast. Such rapid cooling would exert a large strain on natural and societal systems. The probability of this occurring is unknown but presumably much smaller than that of any of the more gradual scenarios included in the Intergovernmental Panel on Climate Change report (Plate 7). The probability is not, however, zero. Obtaining rational estimates of the probability of such a low-probability/high-impact event is crucial. It is worth remembering that models such as those used in the Intergovernmental Panel on Climate Change report consistently underestimate the size and extent of anomalies associated with past changes of the THC; if the underestimate results from lack of model sensitivity possibly linked to overly coarse resolution or other shortcomings rather than from improper specification of forcing, future climate anomalies could be surprisingly large.

Even if no net cooling results from a substantial, abrupt change in the Atlantic THC, the changes in water properties and regional circulation are expected to be large, with possibly large effects on ecosystems, fisheries, and sea level. There are no credible scenarios of these consequences, largely because the models showing abrupt change in the THC have too crude spatial resolution to be used in regional analyses. To develop these scenarios would require the combination of physical and biological models to investigate the effects on ecosystems, and the "nesting" of large-scale and coastal models to investigate sea-level change.

If we are to develop the ability to predict changes in the THC, we must observe its strength and structure as a fundamental requirement, akin to the necessity to observe the equatorial Pacific if one wants to forecast El Niño. So far, however, no observational network exists to observe the THC on a continuous basis. We also need to learn more about which upstream processes and regions are the source of the observed changes in the THC. Present-day observations show substantial decadal changes in the temperature of the warm Atlantic currents flowing toward the Arctic Ocean and in the outflows of freshwater and ice from the Arctic Ocean (e.g., Dickson et al., 2000). Both have been observed to affect the characteristics of the cold deep overflows that cross the Greenland-Scotland Ridge southward to drive the THC (e.g., Hansen et al., 2001b). Systematic, long-term observations of the fluxes influencing the THC are needed. Moreover, remote influences on the THC must be monitored, in particular the low-latitude atmospheric water-vapor transport from the Atlantic to the Pacific and the influence of Southern Ocean changes.

Meltwater from the Greenland ice sheet and glaciers and permafrost also feed into source regions of the THC. Changes in any of these fluxes could contribute to THC changes. The result from Cuffey and Marshall (2000) that modest warming above recent conditions during the previous interglacial led to major shrinkage of the Greenland ice sheet suggests that large changes in Greenland are likely in the future. Changes in river and groundwater discharge are similarly important.

Arctic sea-ice volume appears to have shrunk dramatically in recent decades (Vinnikov et al., 1999; Johannessen et al., 1999; Rothrock et al., 1999). The influence of that decline on the freshwater budget of the Atlantic THC is unknown but could be critical. It is crucial to know the net freshwater flux from the Arctic Ocean to the Nordic Seas, in the form of both sea ice and low-salinity surface water. The sea ice emerging from Fram Strait is thought to influence convection in the Greenland Sea and, after being transported through the Denmark Strait in the East Greenland Current, in the Labrador Sea. Indeed, sea ice from the Arctic could be the origin of such events as the Great Salinity Anomaly that have been documented in the North Atlantic (Dickson et al., 1988; Häkkinen, 1993). Given the importance of freshwater forcing for the stability of the THC, such events might presage change in the circulation. However, even if melting Arctic sea ice does not markedly influence the THC, sea-ice disappearance is likely to have radical consequences for Arctic ecosystems and possibly regional climate. It is not now possible to quantify such possibilities.

5

Economic and Ecological Impacts of Abrupt Climate Change

ost studies of the potential ecological and economic impacts of climate change and greenhouse warming have focused on scenarios that involve gradual climate change. Accumulating evidence from the paleosciences, however, shows that the patterns of change suggested by projections of future climate change are not representative of past climatic changes or of the transitions between different climatic regimes. In particular, many recent paleostudies indicate that current climate is much more stable than were climates in earlier periods.

One notable aspect of large, abrupt global and regional climatic changes is precipitation, which is inherently more variable than temperature. Paleoclimatic records show that extreme and persistent droughts have occurred throughout the last few thousand years in widespread regions, as summarized in Chapter 2. These types of droughts have greatly affected societies. For example, abrupt but persistent droughts have been suggested as the cause of societal disruptions of the Maya (Hodell et al., 1995; Gill, 2000; Figure 5.1). Analogous results have been found for the Tiwanaku cultures (Binford et al., 1997), and Brenner et al. (2001) pointed out that droughts in the Yucatan in the ninth century may be linked across the equator to warming in Peru. Weiss et al. (1993) pointed to the role of abrupt climate shifts in the collapse of third-millennium north Mesopotamian civilization. De Vries (1976) examined the impact of the Little Ice Age in Northern Europe, and there is growing evidence of the impact of droughts and other

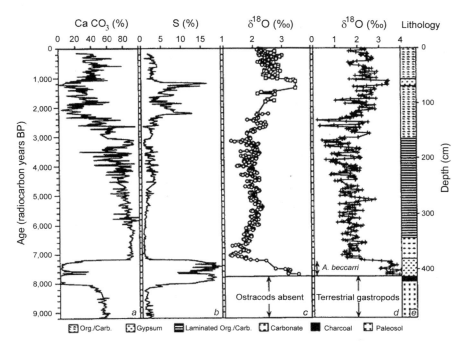

FIGURE 5.1   Core stratigraphy from Lake Chichancanab showing measurements of calcium carbonate, sulphur, and $\delta^{18}O$ of ostracods and the gastropod *P. coronatus*. General lithology is shown at the right of the diagram. Two periods of particularly dry climate (high evaporation/precipitation) are marked by coincident peaks of sulphur content and $\delta^{18}O$ of both ostracods and gastropods. The first occurs between 8,000 and 7,200 years ago, and the second is centered at 1200 during the late Holocene. The latter is linked to the Mayan collapse (from Hodell et al. 1995).

features associated with the El Niño-Southern Oscillation (ENSO) in modern times (e.g., Kitzberger et al., 2001). While the discussion of future climate changes has been dominated by the paradigm of gradual climate warming, we highlight the paleoclimate evidence of Holocene drought because such abrupt changes are likely to be more disruptive of human societies, especially in those parts of the globe that currently have water shortages and marginal rain-fed agriculture.

The recognition of abrupt changes in past climates reinforces concerns about the potential for significant impacts of anthropogenic climate change. Current trends along with forecasts for the next century indicate that the climate averages and variabilities likely will reach levels not seen in instrumental records or in recent geological history. These trends have the potential to push the climate system through a threshold to a new climatic state. What are the likely impacts of anthropogenic changes in climate on human and ecological systems? What kinds of tools do we have to foresee potential impacts? What steps can societies take to reduce and increase adaptiveness?

To gather information and facilitate discussions on these questions, the committee hosted a workshop on the societal and economic impacts of abrupt climate change. The workshop included scientists grouped roughly into the areas of economics and ecology, although this grouping does not adequately capture the broad range of interests and disciplines of the participants. Initially, it may appear that ecology and economics are disparate and unrelated disciplines—seemingly working more at odds than together—with one apparently concerned only with maximizing short-term profits while the other is apparently concerned with maximizing long-term ecosystem preservation. Yet the workshop revealed similarities in methodologies and interests. Both areas are concerned with highly complex, nonlinear, dynamic systems, in which individual firms, industries, plants, or animals interact to produce strange and surprising outcomes. Both disciplines are concerned with empirical understanding of systems and with preserving valuable entities or systems from ill-designed and often inadvertent human interventions. The committee believes that the interaction between workshop participants from these two disciplines was a valuable contribution to the understanding of how abrupt climate change can affect the natural and human-influenced systems. The workshop highlighted a considerable body of work directly relevant to understanding impacts of abrupt climate change exists in disciplines such as archaeology, sociology, and geography, and identified opportunities for the concept of abrupt climate change to motivate better connections between these diverse fields of study.

RECENT SCIENTIFIC STUDIES IN
THE ECOLOGICAL AND SOCIAL SCIENCES

Given the focus of this chapter, it is important to understand the term "abrupt climate change" in the context of ecological and economic systems. As defined in Chapter 1, an abrupt climate change occurs when the climate system is forced to cross some threshold, triggering a transition to a new state at a rate determined by the climate system itself and faster than the cause. Chaotic processes in the climate system may allow the cause of such an abrupt climate change to be undetectably small.

Abrupt climate change generally refers to large-scale events (such as for an entire country or continent), of significant duration (for at least a few years), whose rate of change or variability is significantly greater than the recent variability of climate. From the point of view of societal and ecological impacts and adaptations, abrupt climate change can be viewed as a significant change in climate relative to the accustomed or background climate experienced by the economic or ecological system being subject to the change, having sufficient impacts to make adaptation difficult.

Another important consideration is that there is virtually no research on the economic or ecological impacts of abrupt climate change (Street and Glantz, 2000). Geoscientists are just beginning to accept and adapt to the new paradigm of highly variable climate systems, but this new paradigm has not yet penetrated the impacts community, particularly in economics and the other social sciences. The committee hopes that the diffusion of ideas among the different disciplines investigating climate change happens rapidly, so that research on the impacts of abrupt climate change can move ahead in a timely fashion.

Abrupt Climate Change and Abrupt Impacts

When investigating the impacts of climate change, it is natural to look first for the impacts of abrupt climate changes. An abrupt climate change—whether warming or cooling, wetting or drying—could have lasting and profound impacts on human societies and natural ecosystems. But it must be remembered that profound impacts are not limited to cases of abrupt climate change. Modest changes or increased variability of climate may be sufficient to produce severe impacts, giving the false appearance that these impacts were caused by an abrupt external forcing.

Abrupt impacts result from the fact that economic and ecological processes have adapted to specific climatic patterns and are therefore typically bounded by experience (in the case of society) or history (in the case of ecosystems). Abrupt impacts therefore have the potential to occur when gradual climatic changes push societies or ecosystems across thresholds and lead to profound and potentially irreversible impacts, just as slow geophysical forcing can cross a threshold and trigger an abrupt climate change. Consider that since the nineteenth century, Grand Forks, North Dakota, had successfully fought frequent floods up to a river stage of 49 feet. Then, in 1997, a flood crested at 54 feet and caused catastrophic damages despite the fact that the flood crests were only 10 percent higher than the previous high. This modest difference from typical experience was sufficient to cross an impact threshold (Pielke, 1999).

Research by Pearce (2000) explored impact thresholds for migrating species, describing problems encountered by caribou on their 1,500-km-long trek from winter grounds in the mountains to the Arctic coastal plain in spring. Increased winter snowfall has led to delayed migration and increased river volume. In 1999, snowfall was 50 percent above average, snow melted a month later than usual, and none of the females in the herd made it to the coast before calving. A record low number of calves eventually reached the coast, and some were forced to swim the Porcupine River when only a few days old. These events were observed by the native people in the area, who were moved to reduce their traditional harvest of caribou. The size of the herd dropped from 178,000 in 1989 to 129,000 in 1999. Impacts on the migration of many other species are similarly dependent on boundaries linked to climate.

The Grand Forks floods also help demonstrate the interaction between societal decisions, perceptions of what constitutes "typical climate," and impact thresholds. Following the 1997 Grand Forks floods, the community decided to relocate some properties and build additional levees to raise its threshold to catastrophic impacts. Depending on the assessment of the probabilities and consequences of future flood levels as well as the cost and benefit of flood protection, the community could have chosen 55, 60, or 65 feet as the elevation for the levees. Often, such decisions are made based on assumptions of past weather patterns and runoff. However, if climate is changing, or if the underlying climate system is itself variable, decisions based on past precipitation, runoff, and flood patterns are likely to build in thresholds that incorrectly estimate potential threats compared to decisions based on expectations that allow for changes in climatic means or climate variability. (For more information on the flooding and response in Grand Forks and along the Red River, see International Red River Basin Task Force, 2000.)

Trends in Abrupt Impacts

In the long march of human history, technology has increasingly insulated humans and economic activity from the vagaries of weather. In the preindustrial age, work and recreation were dictated by the cycles of daylight, the seasons, and the agricultural growing season. This was summarized by the economic historian Fernand Braudel, who wrote, "The world (before the nineteenth century) consisted of one vast peasantry where between 80 and 95 percent of people lived from the land and nothing else. The rhythm, quality, and deficiency of harvests ordered all material life" (Braudel, 1973). Gradually, with growing linkages through national and international trade, and as agriculture's share of economic activity has decreased, the role of local weather on harvests (and of climate on the economy) has declined in significance. Many people are surprised to learn that in 1999, farming contributed only $74 billion of the $9.3 trillion (about 0.8 percent) of US gross domestic product. Furthermore, in 1999 agriculture's share of total hours worked also totaled 0.8 percent (Bureau of Economic Analysis, 2001).

Today, modern technology enables humans to live in large numbers in virtually every climate on earth. For the bulk of economic activity, variables such as wages, unionization, labor-force skills, and political factors overwhelm climatic considerations. For example, when a manufacturing firm decides between investing in Hong Kong and Moscow, climate will probably not be on the list of factors considered. Moreover, the processes of economic development and technological change tend progressively to reduce sensitivity to climate as the share of agriculture in output and employment declines and as capital-intensive space heating and cooling, enclosed shopping malls, artificial snow, and accurate weather or hurricane forecasting reduce the vulnerability of economic activity to weather. This trend is seen even in developing countries; countries classified as "low income" by the World Bank (including China and India) had 31 percent of their output coming from agriculture in 1980, while by 1998 that share had declined to 23 percent (World Bank, 2001).

Changes in the historical vulnerability of the US economy to weather can be seen by looking at variability of output in agriculture, which is the most weather-sensitive sector of the economy. The variability is measured as the deviations from trend of real gross output originating in agriculture in 1996 prices over the 1929-2000 period (Figure 5.2)[1] and is caused by a wide variety of factors including weather, floods, exchange-rate changes, demand volatility, as well as bad harvests abroad. The year-to-year variability of agricultural output has risen over time along with the growth in

---

[1]Real gross product output above originating in agriculture is the value added in the agricultural sector, which equals total output less purchases (such as fuel) from other sectors. The quantities (tons of wheat or pounds of bacon) are valued using market prices in 1996.

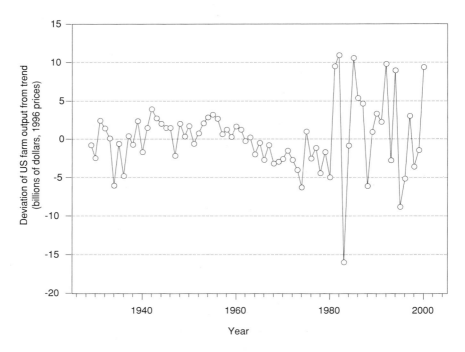

FIGURE 5.2   Variability of total US farm output, 1929-2000. Variability is measured as the deviation of real farm output from its quadratic trend, where output is measured in billions of dollars in 1996 prices. These data include all farm output including cereals, fruits and vegetables, and livestock. (Data from Bureau of Economic Analysis.)

the output of that sector. Surprisingly, however, the vulnerability of the overall economy to agricultural shocks has declined over the last seven decades. The overall vulnerability is here measured as the ratio of the deviation from trend of real gross output shown in Figure 5.2 divided by trend real gross domestic product (Figure 5.3). The maximum deviation due to agriculture over the entire period was –0.75 percent of total output in 1934, while the maximum deviation in the last decade was only 0.14 percent in 1992. The declining sensitivity of overall output to agricultural shocks lies primarily in the declining share of output originating in agriculture. Agricultural output was 5 percent of total gross domestic product in the early 1930s but averaged 1.2 percent of total output in the late 1990s, so weather shocks to farming have a relatively smaller overall impact today than in earlier years.

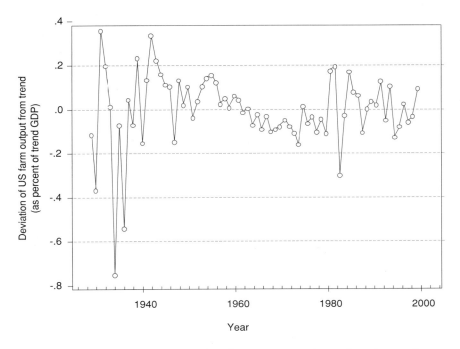

FIGURE 5.3   Relative variability of US farm output as share of total gross domestic product, 1929-2000. Relative variability is the dollar variability shown in Figure 5.2 as a percent of trend real gross domestic product. Even though the variability in Figure 5.2 has increased over the last seven decades, its size relative to the overall economy has declined because of the declining share of agriculture in total output. (Data from Bureau of Economic Analysis.)

Another human vulnerability to climate change is impacts from severe storms. Total economic losses from hurricanes in the United States over the 1900-1995 period have increased sharply (Figure 5.4). In contrast to agriculture, when these data are normalized to account for inflation, wealth, and population (Figure 5.5), they exhibit an extremely skewed distribution but show no clear trend.[2] Furthermore, data on floods show no decline in flood damage per unit wealth in the United States (Pielke and Downton, 2000) over an equivalent period.

The relationship between climate and human civilizations has long been a subject of research and speculation among historians and economists.

---

[2]A least squares estimate of normalized damages on time shows a negative coefficient with an insignificant t-statistic of 0.5.

Millions of 1995 US $

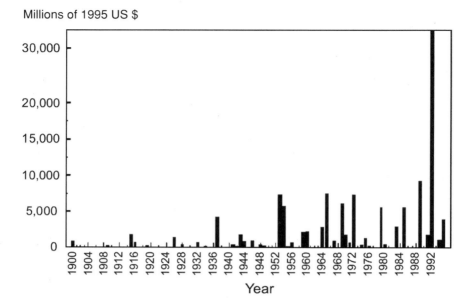

FIGURE 5.4   Annual hurricane damages from 1900-1995. Total economic losses from hurricanes in the United States have increased sharply during this period; however, after accounting for inflation, wealth, and population, no clear trend emerges, as shown in Figure 5.5. (SOURCE: Pielke and Landsea, 1998.)

Much historical work has focused on periods of abrupt climatic changes or politico-economic convulsions. But questions have been raised whether traditional approaches are based on a sufficiently rigorous methodology (de Vries, 1981). Often, studies select extraordinary events (such as revolutions or depopulations) or marginal regions and then look for explanatory events such as climate changes. Unless these crises can be shown to be typical responses to similar changes in climate, the estimated impacts are biased upwards because of statistical selection bias. Such an approach applied to banking would convince people that banking history was essentially the study of bank robberies (de Vries, 1981). To remedy this situation, statistical time-series approaches based on continuous meteorological, economic, and other social data should be used to provide more accurate information regarding the systematic relationship between weather/climate and economic activity (de Vries, 1981).

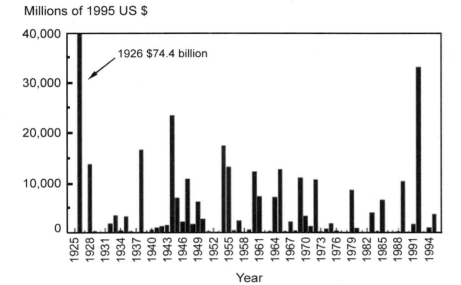

FIGURE 5.5   Normalized hurricane damages from 1925-1995 (adjusting for inflation, coastal county population changes, and changes in wealth). (SOURCE: Pielke and Landsea, 1998.)

Impact of Abrupt Climate Change on Economic and Ecological Stocks

One way of understanding the impact of abrupt climate change is through its impact on economic and ecological stocks or "capital stocks." This approach rests on the idea that most useful economic and ecological activity depends on stocks of capital, where that phrase is defined in the most general sense as accumulations of valuable and durable, tangible and intangible goods and services.[3] In the economy, capital stocks include tangible goods such as factories, equipment, and roads as well as intangible items such as patents, intellectual property, and institutions. Similarly, ecosystems depend on stocks of species, forests, water, and carbon as well as complex "webs" of interacting systems. Viewed along with other inputs,

---

[3]This section draws heavily on Nordhaus (2000). The role of capital stocks is also emphasized in Reilly and Schimmelpfennig (2000).

such as labor and water flows, capital stocks produce most of the world's valuable market and non-market services such as food, recreation, water, erosion control, and many other environmentally related goods and services (sometimes termed "ecosystem services").

Serious impacts to ecological or economic capital stocks can occur when they are disrupted in a manner preventing their timely replacement, repair, or adaptation. It is generally believed that gradual climate change would allow much of the economic capital stocks to roll over without major disruption. By contrast, a significant fraction of these stocks probably would be rendered obsolete if there were abrupt and unanticipated climate change. For example, a rapid sea-level rise could inundate or threaten coastal buildings; abrupt changes in climate, particularly droughts or frosts, could destroy many perennial crops, such as forests, vineyards, or fruit trees; changes in river runoff patterns could reduce the value of river facilities and floodplain properties; warming could make ski resorts less valuable and change the value of recreational capital; and rapid changes in climate could reduce the value of improperly insulated, heated, and cooled houses. There may also be an impact on more intangible investments such as health, technological, and "taste" capital, although these are more speculative.

Similarly, ecological systems are vulnerable to abrupt climate change because they have long-lived natural capital stocks, they are often relatively immobile and migrate slowly, and they do not have the capacity of humans to adapt to or reduce vulnerability to major environmental changes. Ecological systems are also vulnerable because of anthropogenic influences on the environment, which repeatedly alter ecosystems and limit species abundance and composition as a result of habitat disturbance, fragmentation, and loss. Past examples of ecosystem vulnerability to rapid climate change, such as the Younger Dryas cooling, illustrate the fragility of species diversity at one location as forests experienced rapid change. In southern New England, trees such as spruce, fir, and paper birch experienced local extinctions within a period of 50 years at the close of the Younger Dryas (Peteet et al., 1993). North American extinctions of horses, mastodons, mammoths, saber-toothed tigers, and many other animals were greater at this time than at any other extinction event over millions of years (Meltzer and Mead, 1983). The reasons for this extinction have been linked to both climate and early human impacts (Martin, 1984).

The vulnerability of ecological and economic capital stocks to abrupt climate change arises because these stocks are often specific to particular locations and are adapted to particular climates. Demographic rates of migratory songbirds, such as black-throated blue warblers, in both north temperate breeding grounds and tropical winter quarters are shown to vary with fluctuations in ENSO (Sillett et al., 2000). In 1999, warmer waters resulting from El Niño were more deadly to coral reefs (showing a 16 percent loss) than for losses in previous years due to pollution (estimated as a 11 percent loss) (Brown et al., 2000).

The loss in capital value therefore depends on factors such as the lifetime of the capital stocks, the mobility of the capital, the abruptness and predictability of the climate change, as well as the extent to which the capital is managed or unmanaged. For very short-lived produced capital, such as computers or health-care facilities, climate change occurring over two or three decades would have little impact. On the other hand, for dwellings and infrastructure, which have lifetimes of 50 to 100 years, or for slowly adapting and unmanaged ecological systems, such as mature forests, migratory birds, and coral reefs, abrupt climate change could reduce the capital value significantly. The most vulnerable stocks are probably unmanaged ecosystems with long lifetimes ("natural capital"). These capital stocks include forests and similar ecosystems whose lifetimes are on geological time scales, species, and interacting biological systems whose lifetimes are on the biological scale of evolutionary time.

The vulnerability of capital stocks to climate change (measured as the percent of the value destroyed by an abrupt event) is a function of the warning time and the lifetime of the capital stock (Figure 5.6).[4] Suppose, for example, that the owner of capital stock has warning that an abrupt event will occur and it will render the capital completely obsolete. For example, there might be a 20-year warning about shoreline erosion. Under the extreme assumption that the owner can take no remedying measures (such as cost to move the house or abandon the farm), 82 percent of the initial value will be lost for the relatively long-lived capital of 100 years. For capital with relatively short lifetime of 10 years, only 14 percent of the value remains at the end of 20 years, so the vulnerability is relatively low (Figure 5.6).

A detailed simulation of the market value of land and structures of developed coastal properties in the United States on 500 meter by 500-meter samples has been developed to show the dynamics of capital-stock depreciation under different climate-change scenarios (Yohe and

---

[4]Typical lifetimes of capital stocks are 2-3 years for computers, 5-10 years for equipment, 40-80 years for structures, and more than 100 years for some infrastructure (Herman, 2000).

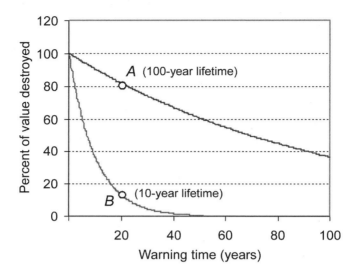

FIGURE 5.6 Vulnerability of capital stocks is higher with reduced warning times and longer lifetimes. The fraction of the value of capital stock that is destroyed or becomes obsolete is a function of the lifetime of the capital stock and the warning time. For an abrupt change that makes the capital completely obsolete, 82 percent of the value is lost for capital with 100-year lifetime and 20 year warning time (A) (top line). For shorter lifetime of 20 years at point B, only 14 percent of the value is lost (bottom line). Note that for a gradual climate change with warning times of at least 100 years, relatively little value is lost even for long-lived capital.

Schlesinger, 1998). This study compared the economic losses from perfectly anticipated and completely unanticipated sea-level rise scenarios ranging from 10 to 90 centimeters over the next 100 years. This example is particularly interesting because the capital stocks involved are long-lived compared to the average reproducible capital stock and coastal structures are among the most vulnerable and immobile capital stocks in our economy.

As would be expected, the results show that the high sea-level-rise scenarios show much higher damages (Table 5.1). Under perfect foresight, the property owner is assumed to optimize the depreciation schedule in light of the need for abandonment when sea-level-rise makes the structure uninhabitable (Table 5.1). Under the myopic case, the owner continues to operate and maintain the dwelling assuming no sea-level-rise until forced to abandon. The ratio of the myopic to the perfect-foresight case ranges from 1.02 for 2050 with slow sea-level-rise to 1.49 for 2100 and the rapid sea-level-rise case (Table 5.1). The interpretation is that, without adaptation, a sea-

TABLE 5.1   Scenarios Comparing the Costs of Sea Level Rise for Different Expectations

	Damages with Perfect Foresight			Damages with Myopic Expectations			Ratio of Myopic to Perfect Foresight		
	10 cm	50 cm	90 cm	10 cm	50 cm	90 cm	10 cm	50 cm	90 cm
Year	[millions of 1990 US dollars per year]						[ratio myopic to perfect foresight]		
2050	8.1	90.6	218.7	8.3	110.4	284.4	1.02	1.22	1.30
2100	14.2	158.3	382.3	16.6	221.8	571.5	1.17	1.40	1.49

NOTE: Two scenarios compare the costs of sea level rise for different expectations. The first set of figures shows the cost of sea level rise for different rates of rise assuming that homeowners and communities perfectly forecast the pace and extent of the rise. The second set of figures shows costs if expectations are myopic, (i.e., assume no sea-level rise). The last set is the ratio of the costs under myopic foresight to perfect foresight. Lack of foresight has higher damages because homeowners make improvements even though the home may soon be damaged or inundated. Modified from Yohe and Schlesinger, 1998.

level-rise of about 1 meter could add about 50 percent to the cost of coastal structures damaged by sea-level-rise (Yohe and Schlesinger, 1998); adaptive capacity is diminished by myopia. This analysis also suggests why ecosystems are particularly vulnerable to abrupt climate change: they tend to be long-lived systems, they are generally unmanaged (e.g., coral reefs) and thus lack the ability to anticipate future events, and their ability to migrate or adapt is slow.

It would be useful to extend an analysis such as that above to the entire economy. This task can be readily undertaken for tangible capital, for which comprehensive valuation and lifetime data exist. But undertaking such an analysis for ecological capital presents a greater challenge because no comprehensive inventory and valuation system exists.

Is Abrupt Climate Change Likely to Exacerbate the Effects of Gradual Climate Change?

Until recently, most research on the impacts of climate change has focused on gradual climate change primarily because the early scenarios developed by climate scientists asked questions such as "what would happen if CO_2 doubled?" and then simulated a situation in which climate moved

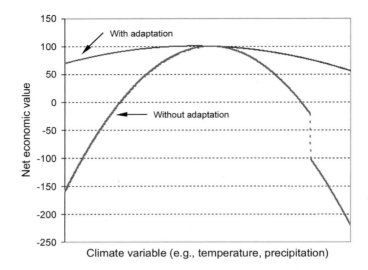

FIGURE 5.7   Illustration of difference between impacts with and without adaptation. The upper line shows the impact of climate change with full adaptation where farmers can change crops and irrigate, where leisure facilities can be replaced, and where forests regenerate. The lower line shows the impacts without adaptation, as is likely to occur with abrupt climate change. Note that for managed systems without externalities, the line without adaptation will lie everywhere below the adaptation line, indicating the costs are likely to be lower with adaptation. We have also shown a break in the no-adaptation line to reflect the potential for sharp threshold effects, such as those due to floods or fire.

smoothly from an initial state to a warmer one. In reality, it is likely that the impacts of abrupt climate change will be both larger and more acute than those under gradual climate change. The major reason for this difference, as discussed previously, is that economies and ecological systems usually will have an easier time adapting to more gradual and better anticipated changes. The net effect of small gradual climatic changes on the economy, as is indicated by most current economic studies for developed countries, is likely to be small relative to the overall economy (Figure 5.7). On the other hand, where economies cannot adapt, impacts resulting from abrupt climate change are likely to be larger and there may be threshold effects if the abrupt climate change triggers fires, water shortages, or exceeds the normal safety margins built into structures or policies.

There is now a substantial literature on the potential impacts of gradual climate change. This was recently reviewed by interdisciplinary teams for the Intergovernmental Panel on Climate Change in its Third Assessment Report (Intergovernmental Panel on Climate Change, 2001a; Ch. 18 is especially relevant). There are also numerous studies of the impacts of climate variability on environment and society at the regional scale, such as caused by El Niño, and some of the more extreme variability considered in these studies may be analogous to abrupt change. There have, however, been few studies that directly analyze the impact of global abrupt climate change. Therefore, the question arises as to whether studies of gradual climate change are useful to assess the impacts of abrupt climate change. In certain areas, existing studies are likely to be instructive. For example, it is likely that those sectors of economies and ecosystems most vulnerable to gradual change are also likely to be vulnerable to abrupt change and, to some extent, the converse should also be true. Agriculture and forests have been identified as systems that are vulnerable to gradual changes, and they are likely to be even more vulnerable to abrupt changes. On the other hand, many manufacturing processes are thought to be relatively robust to climate change at any temporal scale.

For natural systems, it is well documented that gradual climate change can affect species distribution, population abundance, morphology, and behavior, ultimately affecting community structure (Easterling et al., 2000). Although less is known about the links between small- and large-scale climate processes and ecosystems during extremes in climate variability, it is clear that ecosystem structure and function are both affected by disturbances associated with weather extremes such as floods, tornadoes, and tropical storms (Pickett and White, 1985; Walker and Willig, 1999). For example, in the 1950s, widespread drought in New Mexico was responsible for a significant shift of the boundary between ponderosa pine forest and pinyon-juniper woodland (Allen and Breshears, 1998).

In some economic and ecologic sectors, where climatic impacts are of a smooth and linear nature, it is likely that the influence of abrupt climate change will be to accelerate the effects of climate change rather than to qualitatively change the impact. The case of slow sea-level rise discussed earlier provides an example where rapid and unanticipated changes appear to cause only a modest increase in damages. However, under some circumstances, abrupt climate change may not only exacerbate the impacts of gradual climate change but may lead to qualitatively different and more severe impacts.

Abrupt Climate Change Impacts are Likely
When Systems Cross Thresholds

The major impacts of abrupt climate change are most likely to occur when economic or ecological systems cross important thresholds. For example, high water in a river will pose few problems until the water runs over the bank, after which levees can be breached and massive flooding can occur. Many biological processes undergo shifts at particular thresholds of temperature and precipitation (Precht et al., 1973; Easterling et al., 2000). For example, many plants are adapted to a specific climate and limited by parameters such as frost and drought tolerance (Woodward, 1987; Andrewartha and Birch, 1954). Phenology, the progression of biological events throughout the year, is linked to climate. Several studies have demonstrated this link for climatic warmings over the last half century. For example, in southern Wisconsin a phenological study of over 61 years of springtime events showed that several of these events were occurring earlier in the year. The events that did not are probably linked to photoperiod or regulated by a physiological signal other than temperature (Bradley, 1999; Leopold, 2001). Data spanning 60 years in Britain show that breeding patterns of birds are linked to the North Atlantic Oscillation (NAO), with 53 percent of birds showing long-term, significant trends toward earlier breeding since 1939 (Crick et al., 1997; Crick and Sparks, 1999). The trends toward earlier nesting in birds are paralleled in studies of bird migration (Sparks et al., 1999), butterfly emergence (Roy and Sparks, 2000), and flowering (Schwartz and Reiter, 2000). In North America, similar studies have noted earlier egg laying in tree swallows (Dunn and Winkler, 1999), earlier robin migration, and earlier exit from hibernation by marmots (Inouye et al., 2000). Although these studies indicate a change in organism behavior relating to a steady shift in climate over time, abrupt shifts may exceed the limits of organisms' ability to adapt.

Many important threshold effects occur at the boundaries of systems. Ecotones, the narrow zones where ecological communities overlap, are particularly susceptible to abrupt climate change, primarily because the species diversity is great and the vegetation is often limited by a sharp climatic gradient. Paleorecords of the decadal response of forest dieback (Peteet, 2000) demonstrate how rapidly boreal forest can be replaced by mixed hardwoods, as was observed in eastern United States ecotonal forests at the close of the Younger Dryas. European pollen records from the cold event about 8,200 years ago indicate significant species changes in fewer than 20 years (Tinner and Lotter, 2001). Furthermore, modern studies also point to climatic stress as a probable cause of many historical forest declines (Hepting, 1963; Manion, 1991). For example, a few unusually warm summers were associated with past declines of red spruce (*Picea rubens*) in eastern North America (Cook and Johnson, 1989).

Extremes of environment are most damaging to the reproductive stages of plants. For example, changes in mast fruiting,[5] which are often synchronous over large regions, have strong effects that cascade through all levels of an ecosystem (Koenig and Knops, 2000). One example is the influence that large acorn crops have on increasing the populations of deer, mice, and ultimately ticks (Jones et al., 1998). Thus, climatically induced changes in masting that lead to increased acorn production can result in an enhanced risk of Lyme disease, which then impacts human health. It is likely that the effects of abrupt climate change on mast fruiting will be nonlinear and thus the impacts of these changes will be difficult to predict (Koenig and Knops, 2000).

Drought is also of primary importance to forests. In contrast to earlier predictions that global warming would increase radial growth of trees in boreal forests, white spruce (*Picea glauca*) tree ring records show recent decreases in radial growth. These decreases are presumed to be due to temperature-induced drought stress, which has implications for forest carbon storage at high latitudes. In the Southern Hemisphere (Patagonia), recent pulses of mortality in *Austrocedrus chilensis* trees were associated with only 2 to 3 years of drought (Villalba and Veblen, 1998). Not only is the lack of water directly damaging in a drought, but there is increased susceptibility to fire as a forest dries out. Further, there is evidence that drought triggered an ecotonal shift in New Mexico (Allen and Breshears, 1998) where ponderosa pine experienced high mortality rates in less than 5 years and the ecotone migrated over 2 km. Woody mortality loss occurs much faster than tree growth gain, which has pervasive and persistent ecological effects on associated plant and animal communities.

Ice-core records from Greenland record changes in the frequency of layers enriched in fallout from forest fires (Taylor et al., 1996). The frequency of occurrence of layers was anomolously high during the abrupt cold event about 8,200 years ago, which in turn is associated with drought in regions upwind of Greenland, likely including the United States midcontinent (Alley et al., 1997). In the southwestern US, synchrony of firefree and severe-fire years across diverse forests implies that climate forces

---

[5]Mast fruiting is the production of seeds by a population of plants.

fire regimes on a subcontinental scale (Swetnam and Betancourt, 1990; Plate 9).

Droughts and floods are also responsible for changes in erosion patterns, as reduced vegetation due to fires results in greater soil loss (Allen, 2001). For example, in the Indonesian tropics, drought years have led to a greater frequency and magnitude of fires resulting in a loss of peatlands, increased erosion, and increased global air pollution. Globally, rates of soil erosion are 10 to 40 times greater than rates of soil formation (i.e., over 75 billion tons from terrestrial systems annually; Pimentel and Kounang, 1998).

Droughts have also been implicated in insect outbreaks and pulses (births and deaths), with impacts on tree demography (Swetnam and Betancourt, 1998). Episodic outbreaks of pandora moth (*Coloradia pandora*), a forest insect that defoliates ponderosa pine (*Pinus ponderosa*) and other western US pine species, have been linked to climatic oscillations (Speer et al., 2001). Drought years have been linked to insect crashes as well as booms (Hawkins and Holyoak, 1998).

Tundra systems are highly susceptible to the effects of climate change because of their sensitivity to water table fluctuations, snow-albedo feedbacks, fire frequency, and permafrost melting (Gorham, 1991). Tundra areas are significant in the terrestrial/atmospheric carbon balance because tundra peat is a large source and sink of greenhouse gases, most notably carbon dioxide and methane (Billings, 1987). Furthermore, 15 to 33 percent of the global soil carbon is contained in the northern wetlands (Gorham, 1991; Schlesinger, 1991). The sensitivity of tundra ecosystems to abrupt climate change, and the widespread influence these areas have on feedbacks in the climate system, make these key areas for climate change monitoring. For example, field measurements and modeling have shown that even vegetation changes induced by summer warming could result in climatic feedbacks (Chapin et al., 2000).

Paleoclimatic investigations have shown that changes in biodiversity are correlated to climate variations. Abrupt climate shifts have been linked to increases in biodiversity as well as to species extinction, although extinctions can occur much more rapidly than can origin of diversity by evolution. The evolution of large beaks among Darwin's finches (*Geospiza fortis*) is thought to be the result of ecological stress from El Niño drought years, while extremely wet years favored evolution of small beaks (Boag and Grant, 1984). But recent research suggests that the rate of species extinctions is on the order of 100 to 1,000 times higher than before humans were dominant (Lawton and May, 1995; Vitousek et al., 1997). For example, over the last two millennia, one-fourth of all bird species are believed to have gone extinct as a result of human activities (Olsen, 1989). This increase in species extinctions is in part because extinctions are occurring on the time scale of human economies, but evolution typically occurs more slowly (except for microorganisms, including those that cause disease).

Despite convincing scientific evidence, there seems to be a general lack of public awareness concerning the global decline in biodiversity, related ecological and societal impacts, and the fact that biodiversity changes are not amenable to mitigation after they occur (Chapin et al., 2000) (Figure 5.8). Overall, land transformation is identified as the primary driving force in the loss of biodiversity worldwide (Vitousek et al., 1997). It is estimated that 39 to 50 percent of available land has been degraded by human activity (Vitousek et al., 1986; Kates et al., 1990). Human influences include loss of permeable surfaces to pavement and increases in greenhouse gases, aerosols, and pollutants. Humans use over half the world's accessible surface fresh water, and more nitrogen is fixed by humans than by all natural sources combined (Vitousek et al., 1997). After land transformation, the next most important cause of extinctions is invasion of nonnative species. Many biological invasions are irreversible, degrade human health, and cause large economic losses. The impact of the zebra mussel on the US Great Lakes states is one such example.

Montane ecosystems are valued highly even though they are small in area compared to other major biomes. Recognized by scientists for their importance to hydrological, biochemical, and atmospheric processes, they are particularly sensitive to climate change because of their many ecotones (Spear et al., 1994) and limited spatial area. Montane areas are scenic, relatively pristine, and often provide water for many urban and agricultural areas (Walker et al., 1993).

Changes in snowfall and snowpack in montane regions have a large impact on treeline and plant communities (Patten and Knight, 1994). The timing of snowmelt affects plants in many ways, including the alteration of leaf traits, the changing of leaf production, shoot growth (Kudo et al., 1999), and flowering (Inouye and McGuire, 1991). In addition, changes to the timing of snowmelt will alter the exposure to late spring frosts (Inouye, 2000). A decrease in synchrony associated with phenological events at high and low altitudes may pose problems for animal species that migrate between altitudinal zones (Inouye et al., 2000). For example, marmots are presently emerging from hibernation 38 days earlier than they did 23 years ago in the Rocky Mountains (Inouye et al., 2000). As this example relates

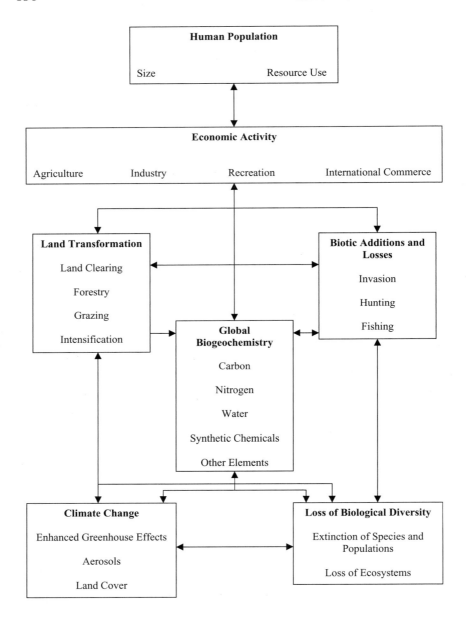

FIGURE 5.8   The many direct and indirect impacts of humanity on the earth's natural ecosystems is shown, with links between climate change, enhanced greenhouse gases, and biological diversity (modified from Vitousek et al., 1997).

to gradual climate change, the impacts of abrupt climate change are expected to be greater.

Wetlands are among the most important ecosystems on earth, due to their many "ecosystem services." They serve as "kidneys of the landscape" by cleansing polluted waters, recharging aquifers, and protecting shorelines (Mitsch and Gosselink, 2000). Wetlands, including estuaries, provide protection for organisms and serve as nurseries for the young of many oceanic food chain species. Furthermore, wetlands can be important carbon sinks. Only recently valued for stabilization of water supplies, wetlands have been filled, drained, or ditched throughout history. In the United States, over half of wetland areas have been drained (Mitch and Gosselink, 2000).

Wetlands are particularly vulnerable to abrupt climate change, especially droughts. Along with lakes and rivers, wetlands are dependent on precipitation for the health of their associated ecosystems. The paleorecord of droughts recorded by wetlands provides an indication of the sensitivity of these areas to climate change. These records are useful for predicting the frequency of abrupt climate change and the sensitivity of water bodies to future climate change. The impacts of abrupt climate changes will be exacerbated because of alteration to the natural environment from the impoundment of rivers, depletion of fossil groundwater, and draining of wetlands.

Montane glaciers are among the most responsive indicators of climate change. Past records of glacial advance and retreat are important for understanding relative magnitudes of abrupt climate change. These records show that the global altitude drop of the equilibrium snow line during abrupt climate changes was of similar magnitude at widely scattered locations, implying that an overall cooling of the atmosphere took place, not a redistribution of the heat balance (Broecker and Denton, 1989; Lowell, 2000; Broecker, 2001). Loss of montane glacier volume has been more or less continuous since the nineteenth century, but loss in the last decade has accelerated (Dyurgerov and Meier, 2000). The loss of montane glaciers in the tropics is dramatic, and in a warmer world this loss may extend to many temperate locations used for recreation.

Ice sheets are linked to abrupt climate change because melting of Greenland or the West Antarctic ice sheet would add directly to global sea level rise and to possible changes in the thermohaline circulation (Manabe and Stouffer, 1997). Much attention has been focused on the possibility of a rapid collapse of the West Antarctic ice sheet. Recent geological and glaciological evidence points to a stable but net decay since the last ice age (Conway et al., 1999), but with considerable uncertainty about future trends and the possibility of rapid dynamic response to future warming. The Greenland ice sheet has the potential for rapid surface melting and perhaps enhanced ice flow with continued greenhouse warming. Laser-altimeter surveys in the 1990s indicated an overall negative mass balance for Greenland ice that results in a 0.13 mm per year sea level rise (Krabill et al., 2000). Since the late 1800s the margin of the Greenland ice sheet has retreated 2 km in some places (Funder and Weidick, 1991) indicating that Greenland ice is responding to twentieth century warming. The influence of the Greenland ice sheet system on potential abrupt climate change appears to be linear except for the possibility of threshold changes in ocean circulation, but the existence of dynamically controlled ice streaming at least suggests the possibility of dynamical changes (Fahnestock et al., 1993).

Much attention has focused on observations that indicate a reduction in the extent and thickness of Arctic sea ice (Cavalieri et al., 2000; Rothrock et al., 1999; Wadhams and Davis, 2000; Parkinson, 2000). Sea ice trends are less clear in the Antarctic. Published estimates from satellite measurements include no significant trend in sea ice from 1973-1996 (Jacka and Budd, 1998), and over a slightly different interval, an increase of 1.3 ± 0.2 percent per decade in mean-annual sea-ice area from 1978-1996 (Cavalieri et al., 2000). However, an interpretation of whaling records indicates a large and rapid decrease in Antarctic summertime sea ice extent between the mid 1950s and early 1970s (de la Mare, 1997). This variability, whether natural or human-induced, is important because it is large and may contribute to "surprises" that may accompany future climate change (Broecker, 1987).

Similar concerns regarding the impacts of abrupt climate change on thresholds apply to social systems. Over human history, one of the major ways humans have adapted to changing economic fortunes has been to migrate from unproductive or impacted regions to more productive and hospitable regions. Until the twentieth century, national boundaries were often open, allowing people to migrate freely in response to economic conditions. For example, as a result of the potato blight (*Phytophthora infestans*) in Ireland, there was a disastrous famine between 1845 and 1847 and almost one million people emigrated, mainly to America. Because national borders may be less open today, it may be difficult for people to migrate to other countries when famines or civil wars occur. These "boundary effects" could be particularly severe for small and poor countries, whose populations are often unwelcome in richer countries. To the extent that abrupt climate change may cause rapid and extensive changes of fortune for those who live off the land, the inability to migrate may remove one of the major safety nets for distressed people.

SECTORAL APPROACHES

One promising approach to investigating the economic impacts of abrupt climate change would be to focus on individual sectors. Economic studies often have relied on the organizing framework of the national income and product accounts, in which the economy is divided into major sectors and for which both income and output data are available for most countries and numerous years (see Toth, 2000). For example, for the United States, detailed income and output data are available for 83 different sectoral breakdowns during the period 1948-1999. There is no similar accounting framework for nonmarket sectors or for ecological studies.

Agriculture

Under the sectoral approach, there are a few obvious sectors to examine. Perhaps the most important is agriculture, which is the sector most heavily affected by weather and climate. Current studies indicate that the impact of gradual climate change on agriculture over the next few decades is likely to be relatively modest for the United States. For example, a recent survey found that the impact of a 2.5°C gradual warming is likely to reduce agricultural value between 0.1-0.2 percent of global income (Nordhaus and Boyer, 2000).

Agriculture is likely to be the most accessible sector for studies on the impact of abrupt climate change because of the extensive information available and the well-developed research infrastructure in this field (Reilly, submitted). It is important to emphasize that it will be necessary to view different abrupt climate impact scenarios from a probabilistic rather than a mechanistic point of view. This applies not only to agriculture but also to other areas such as species extinction, floods, wildfires, hurricanes, human and wildlife diseases, and droughts. It is unlikely that an abrupt climate change will occur as a single abrupt event; rather, it will occur as a distribution of potential events, with increasing severity at lower probabilities. This probabilistic nature of abrupt climate change adds another layer of complexity to impact analyses and to communicating the risks and opportunities to the public.

Because of a methodological coincidence, certain approaches used to analyze the effects of gradual climate change on agriculture can be applied for abrupt climate change. Many early studies of the impact of gradual climate change on agriculture used what came to be referred to as the "dumb-farmer" scenario. Under this approach, farmers were assumed to take only the most limited steps to adapt to climate change. For the "dumbest" of scenarios, even with a 3-6°C warming over a century, farmers would plant the same crops in the same place with the same fertilizers and the same planting and harvesting dates. These assumptions were criticized by those advocating a "smart-farmer" world. Increasingly, studies now allow for extensive adaptation, including the "Ricardian approach" (relying on cross-sectional data on land prices; Mendelsohn et al., 1994), which assumes virtually complete adaptation to climate change.

Ironically, studies that were undertaken using a non-adaptation approach are ideal to investigate the impacts of abrupt climate change. This is particularly true if the abrupt climate change is unforeseen, widespread, and sudden, for in such a case farmers could only undertake very limited adaptation. With full adaptation, linked economic-climate models project essentially no impact on the economic value of global agriculture even for globally averaged temperature changes of as much as 4°C (Figure 5.9).[6] With very limited adaptation, as is likely to occur with abrupt climate change, the costs are substantial, ranging from $100 billion to $250 billion depending on the climate scenario (Figure 5.9). Moreover, these scenarios are likely to underestimate the damages from abrupt climate change because they predict relatively smooth regional changes, whereas evidence suggests that abrupt climate change will produce abrupt, magnified changes at local scales. Note that this figure shows exactly the kind of pattern of impact that was illustrated hypothetically in Figure 5.9.

Forests

A second priority sector for research is forests. This sector may be affected more than agriculture by abrupt climate change because forests are highly climate-sensitive, with long-lived ecological and economic capital stocks. Most ecological studies of the impact of climate change on forests have been limited to describing long-term equilibrium effects (Mendelsohn, 2001). The models predict ecological changes that will occur in two or

---

[6]Note that these scenarios have regionally differentiated climate changes, although transitions between the changes are generally quite smooth.

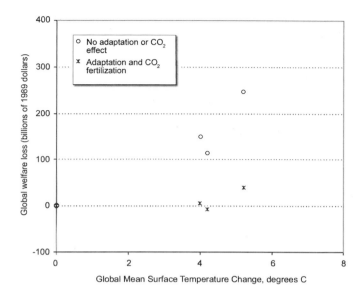

FIGURE 5.9   Illustration of the impacts of gradual climate change with adaptation and abrupt climate change without adaptation (Reilly, submitted). Global welfare losses from agricultural impacts for different climate and CO_2 fertilization scenarios. Upper plots (o) represent global welfare loss associated with abrupt climate change as modeled by an assumption of no adaptation and no CO_2 fertilization. The lower plots (x) allow for full adaptation and include CO_2 fertilization. The three different results are three alternative global yield scenarios from Rosenzweig and Parry (1994) as simulated by Reilly, Hohmann, and Kane (1993) or briefly reported in Reilly and Hohmann (1993). Note that all scenarios go through the origin as a normalization of no welfare loss with no climate change.

more centuries after climate settles down to a new level. Ecologists disagree on how quickly ecological systems will adjust and what they will look like in the process of adjustment. Some ecologists predict vast and rapid losses of forests and grasslands to fires and insects. Others predict slow and gradual change as reproductive competition dominates the dynamics. Both may be correct in different locations, but it seems safe to conclude that there is little secure knowledge about how these new ecosystems will react to climate change.

Abrupt climate change would increase the required adjustment and the uncertainty for forests. Ecosystems may collapse more rapidly, with forests disappearing in vast fires and grasslands dying and turning lands into dust bowls. Such events have occurred in historical times, earlier during the

Holocene, and probably occurred during the glacial periods when dust levels were significantly higher around the world (Overpeck et al., 1996). Ecosystems may be dominated by early successional species. Given that the economy depends on many late-successional species, losses of existing trees would dig into inventories of standing timber and raise timber prices, perhaps for long periods of time. Wildlife dependent on mature systems may be especially stressed; landscape appearances may dramatically change; and runoff may change completely from historic ranges. These changes suggest the potential for large damages.

The impact of climate change on forest systems is currently an area of intensive research (e.g., Hansen et al., 2001a; Shafer et al., 2001), although the focus is on gradual climate change. These studies suggest that there may be some immediate damages associated with dieback in forests and that long-term productivity may increase in some species if CO_2 fertilization occurs, but not in others. Some paradoxes lurk here, however. In a study of how forests in the United States might adjust to future climate change, Sohngen and Mendelsohn (1998) were surprised to find that a scenario with significant dieback of existing stands did almost as well as a scenario where natural change occurred more slowly (Sohngen and Mendelsohn, 1998). Their study illustrates the important point that adaptation (or well-designed management) can help ecosystems to adjust more rapidly and with lower overall economic costs.

Little is known about how climate-change impacts on forests would affect nonmarket services. Little is known about what would happen to wildlife because of complex interactions, and how much people would value the changes. Forest products such as fruits, nuts, medicine, and mushrooms are highly valued, but it is not clear how the flow or value of these products might change with abrupt climate change. Although ecologists predict that biomes will shift with warming, there has been little accompanying social science research to evaluate such shifts. There has been scant research on how forests might change in appearance or how much people would care about the changes. Thus, many of the quality-of-life effects on forests have not been evaluated.

Water Resources

There is a growing awareness of the scarcity of freshwater at the earth's surface compared with the total amount of water on the planet, and concern about how climate change might affect water resources (Gleick, 2000;

Postel, 2000; Sampat, 2000). According to these authors, only 3 percent of the earth's water is fresh, and most of this water is tied up in glaciers and ice sheets. Only about 0.01 percent of the total water is available in lakes, the atmosphere, and soil moisture. Water that is "clean" in terms of its suitability for human use is estimated today to be unavailable for 20 percent of the world's population. Humans have a long history of arguments and conflict over water rights. Water from rivers is impounded for energy generation, agriculture, dilution of wastes, and transport. The construction of dams affects biotic habitats both directly and indirectly; for example, damming of the Danube River has altered the silica chemistry of the entire Black Sea (Vitousek et al., 1997). In the United States only 2 percent of rivers flow unimpeded, and the flow is regulated in over two-thirds of all rivers (Abramowitz, 1996).

The water cycle has important nonlinearities that are relevant to the impacts of abrupt climate change (Kling, 2001). First, the water balance of lakes is extremely sensitive to climate change, and small shifts in the ratio of precipitation to evaporation can result in large changes in lake levels. Extremely large shifts in paleolake levels have been tied to abrupt climate change that is global in extent (Broecker et al., 1998). Lake levels in Lake Victoria and throughout Africa rose suddenly at the beginning of the Bolling warming about 14,700 years ago, a warming that is evident in Greenland, Europe, and North and South America. During the Holocene, rapid lake level shifts are found in western North America (Laird et al., 1996; Stine, 1994) as well as southern South America and Africa. Changes in the Great Lakes water levels have also been linked to climate change, most recently in the last few decades (Sellinger and Quinn, 1999). The declining extent of the Aral Sea and Lake Chad involve both human impact and climate change (Kling, 2001).

Second, the depletion of groundwater is highly nonlinear in its impact on water use when the aquifers dry up (i.e., going from a situation of plentiful water to one of no water or very salty water over a period of a few months). Groundwater deficits, which represent the amount of water withdrawn compared to the amount of recharge water put back into the aquifer, are now equal to at least 163 cubic kilometers per year throughout the world, which is equivalent to about 10 percent of the world's grain production (Brown et al., 2000). In many cases groundwater is nonrenewable, or fossil water, that took extremely long periods to accumulate and is now being used more rapidly than it is replenished. For example, three-quarters of the water supply of Saudi Arabia currently comes from fossil water

(Gornitz et al., 1997). Finally, pollution of existing groundwater supplies is a persistent problem in many regions. Understanding the causes and consequences of these deficits and problems is critical for managing the future of water resources on earth (Kling, 2001).

Third, a nonlinear response exists between soil moisture and the production of greenhouse gases. As soil dries, microbes limited by water respire less soil organic matter, which produces less carbon dioxide and methane. As soil moisture increases, respiration increases and more greenhouse gases are produced. But if the soils are waterlogged, they become anoxic and overall microbial activity is again reduced (leading to increased net carbon storage). Thus knowing where any system lies on this nonlinear response curve is necessary to predict how greenhouse gas production will change as water availability changes (Kling, 2001).

Fourth, many biological systems are nonlinear in their response to water. Outbreaks of vector-borne diseases, such as cholera outbreaks linked to El Nino and water-temperature changes (Colwell, 1996; Pascual et al., 2000) and other unpleasant events are often linked to deviations in precipitation patterns (Balbus and Wilson, 2000). Both extremes in precipitation can be harmful, because some outbreaks are linked to precipitation increases and flooding, while others are associated with droughts. Both extremes lead to stresses on organisms and ecosystems, and as such the biological nonlinear and complicated response may have undesirable effects for humans.

There is virtually no direct research on the economics of water systems and abrupt climate change at this time (exceptions are Glantz, 1999; Arnell, 2000). Given the importance of water for many aspects of society, from agriculture to electricity production, as well as the documented record of periodic droughts, this should be one of the priority areas for research.

Human and Animal Health

Yet another area of potential concern is health—both of humans, of domesticated plants and animals, and of wildlife (National Research Council, 1999a). There is widespread appreciation of the potential for unwelcome invasions of new or exotic diseases in the human population, particularly of vector-borne diseases such as malaria. Similar concerns may arise for pests and diseases that attack livestock or agriculture. Another concern is diseases of wildlife.

Scenarios based on climate models for greenhouse warming indicate that changes will occur in the geographic distribution of a number of waterborne diseases (e.g., cholera, schistosomiasis) and vector-borne diseases (e.g., malaria, yellow fever, dengue, leishmaniasis) if not countered by changes in adaptation, public health, or treatment availability. These changes will be driven largely by increases in precipitation leading to favorable habitat availability for vectors, intermediate and reservoir hosts, and/ or warming that leads to expansion of ranges in low latitudes, oceans, or montane regions. The host-parasite dynamics for abrupt climate change have not been targeted specifically as yet, but Daszak et al. (2001) suggested three phenomena that indicate abrupt climate change may have had heightened impacts on key human diseases:

1. There appears to be a strong link between El Niño-Southern Oscillation (ENSO) and outbreaks of Rift Valley fever, cholera, hantavirus, and a range of emergent diseases (Colwell, 1996; Bouma and Dye, 1997; Linthicum et al., 1999), and if ENSO cycles become more intense, these events may become more extensive and have greater impact;

2. Malaria has reemerged in a number of upland tropical regions (Epstein, 1998) (although this is debated by Reiter, 1998); and

3. Recent extreme weather events have precipitated a number of disease outbreaks (Epstein, 1998).

Criteria that define emerging infectious diseases of humans were recently used to also identify a range of emerging infectious diseases that affect wildlife (Daszak et al., 2000). They include a fungal disease that is responsible for mass mortality of amphibians on a global scale and linked to species extinctions (Berger et al., 1998), canine distemper virus in African wild dogs, American ferrets and a series of marine mammals, and brucellosis in bison as well as others. An ongoing reduction in biodiversity and increased threats of disease emergence in humans and livestock make the impacts of these changes potentially very large.

Emerging diseases are affected by anthropogenic environmental changes that increase transmission rates to certain populations and select for pathogens adapted to these new conditions. Daszak (2001) points to abrupt climate change as pushing environmental conditions past thresholds that allow diseases to become established following their introduction. For example, African horse sickness (a vector-borne disease of horses, dogs, and zebras) is endemic in sub-Saharan Africa. Although it usually dies out within 2 to 3 years of introduction to Europe, the latest event involving imported zebras to Spain resulted in a 5-year persistence, probably because recent climate changes have allowed the biting midge vector to persist in the region (Mellor and Boorman, 1995).

Introduced diseases are costly—a single case of domestic rabies in New Hampshire led to treatment of over 150 people at a cost of $1.1 million. The cost of introduced diseases to humans, livestock, and crop plant health is estimated today at over $41 billion per year (Daszak et al., 2000). Abrupt climate change-driven disease emergence will significantly increase this burden. Furthermore, the economic implications of biodiversity loss due to abrupt climate change-related disease events may be severe, as wildlife supports many areas (fisheries, recreation, wild crops) very significant to our well-being.

Other Sectors and Technological Responses

Work in other sectors is still in its infancy. The results for sea-level rise (without storms) were reviewed earlier in the chapter, and the preliminary results reported there indicate that the cost of unforeseen climate change was only modestly more expensive than anticipated climate change. These estimates would change if storms were also included (West et al., 2001), emphasizing that work will need to consider variability and extremes as well as trends.

In addition, little attention has been given to the impact of abrupt climate change on leisure and recreational activities such as tourism. Such attention may seem frivolous, but in fact a substantial part of human time and of economic output is devoted to outdoor recreation—fishing, hunting, bird-watching, golfing, skiing, swimming, walking, and so forth. Efforts to understand the impact of climate change on leisure and recreation are hampered by lack of systematic data on time use of the population.

It is useful to remind ourselves that over the longer term, technological changes can modify the impact of climate and of abrupt climate change on human activities, on ecosystems, and on economic welfare. Irrigation and fertilization change the impact of climate on agriculture; plant and animal breeding have produced drought- and pest-resistant varieties; forest management practices have changed the pattern of tree growth and timber harvesting; land fills, dikes, and sea walls have kept the sea at bay in low-lying regions; dams and drip irrigation have spread out available water supplies; heat pumps and white roofs can change local energy balances; social and private insurance have reduced individual vulnerability to extreme weather events; vaccines and medications have reduced the impact of many diseases that earlier rendered the tropics uninhabitable; snow-making equipment has turned marginal mountains into thriving ski areas. While few of these are unalloyed improvements, and many have unwelcome impacts on ecosystems, they have reduced human vulnerabilities markedly. Moreover, it is unlikely that these trends will suddenly stop tomorrow. Thus, one of our societal challenges is to harness technology to improve monitoring and prediction and enhance adaptation in the face of potential abrupt climate change. A second challenge is to better understand the interactions and feedbacks between the natural and human-dominated systems, and the limitations of technology when dealing with ecosystem vulnerabilities.

MODELING THE IMPACTS OF CLIMATE CHANGE

Economic and ecological systems are extremely complex nonlinear systems. Both involve individual components (whether firms or organisms) which interact in ways where "everything depends on everything else." Because of this complexity, economists and ecologists increasingly have turned to computerized numerical models to help understand the functioning of these large systems and to help predict the impacts of shocks or disturbances.

Over the last decade, these models have been deployed to help understand the potential impacts of climate change on economic and ecological systems. As a result, substantial advances have been made in understanding the way that economic and ecological systems may be affected by gradual climate change. One of the important developments in economics over the last decade has been the development of integrated-assessment economic models, which analyze the problem of global warming from an economic point of view. Integrated-assessment economic models are empirical computerized models that incorporate the major linkages between climate change and the economy. Dozens of modeling groups around the world have brought to bear the tools of economics, mathematical modeling, optimization, decision theory, and related disciplines to investigate the effects of accumulations of greenhouse gases as well as the costs and benefits of potential policy interventions such as emissions limitations or carbon taxes.[7]

As with economic impact studies, virtually all the studies undertaken by the integrated-assessment economic models community of scholars have

---

[7]A survey of recent developments in integrated assessment economic models is contained in Weyant et al. (1996), Parson and Fisher-Vanden (1997), and Toth et al. (2001).

addressed the questions of gradual climate change. Recently, a few researchers have investigated the impact of threshold events, such as a weakening and eventual shutdown of the North Atlantic thermohaline circulation (for shorthand, such events are called thermohaline circulation collapse or THCC). One set of studies has examined "inverse" analyses in which emissions trajectories were constructed that would keep greenhouse gas accumulations away from the threshold at which a THCC would occur (Toth, 2000). A further set of studies examined policies that would keep climate short of a "catastrophic" threshold (Nordhaus, 1994; Nordhaus and Boyer, 2000).

One recent set of studies (Keller, 2000; Keller et al., 2001) is particularly relevant because it shows that threshold effects can have surprising impacts on efficient economic policies. These studies begin with the dynamic integrated model of climate and the economy model of climate change (Nordhaus, 1994; Nordhaus and Boyer, 2000) and introduce an ocean-circulation model that allows for a THCC. They assume that the THCC was essentially irreversible and would impose a large economic penalty (either in the sense of actual economic losses or in the sense of a large willingness to pay to prevent the THCC). They then couple the dynamic integrated model of climate and the economy model of climate change with their ocean circulation model to investigate the efficient economic policy. An "efficient" policy is one that minimizes the net economic impact or maximizes net incomes when taking into account both the damages from climate change as well as the costs of slowing climate change.

Figure 5.10 shows the result of the Keller et al. (2001) calculations. This shows the abatement rate (measured as a percent of baseline emissions reduced) as a function of time on the horizontal axis and the damage on the vertical axis. Low abatement rates occur early and for low THCC costs, while high abatement rates occur in later periods and for high THCC costs. Two features of the result are worth noting. First, the abatement rate or carbon tax is relatively insensitive to the cost of the THCC for the first half of the twenty-first century. The reason is that it is economically advantageous to postpone costly abatement because of such factors as discounting due to the productivity of investments or because society might discover an economical low-carbon fuel. A second and surprising result occurs for damage cost in the range of 0.5 to 1.5 percent of gross world product. For this region, abatement rises rapidly until around 2100 and then actually declines. The reason is that the early high abatement delays the THCC for a few years and thereby postpones the costs; however, once reversal has oc-

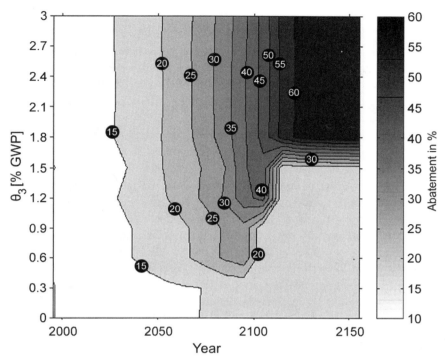

FIGURE 5.10   Efficient abatement policies as a function of threshold specific damages (Keller, 2000). The efficient carbon abatement policies are derived in an "optimal economic growth model." The figure shows abatement rate contours (reduction rate as a percent of baseline) as a function of time on the horizontal axis and damages from a shutdown of the thermohaline circulation as a percent of world output on the vertical axis. The economic model is based on Nordhaus (1994), modified to allow for an ocean thermohaline circulation collapse. Ocean modeling studies suggest that the thermohaline circulation may shut down when the equivalent CO_2 concentration rises above a critical value (Stocker and Schmittner, 1997). The model assumes a CO_2-temperature sensitivity of 3.6°C for CO_2 doubling and neglects parameter uncertainty. (See Keller, 2000 and Keller et al., 2001 for details.)

curred, then abatement only occurs from the non-THCC part of climate damages.

These first efforts to apply integrated-assessment economic models to abrupt climate change are useful in suggesting how abrupt climate change might influence our thinking about the nature of policies to slow climate change. At the same time, attention needs to be paid to other shocks. In one sense, the THCC is an extremely nasty climatic outcome but, as modeled to date, not really abrupt. It would be useful to pay more explicit attention in integrated-assessment economic models to other kinds of potential shocks, perhaps focusing on major changes in water and agriculture systems for the United States as well as to shifts in monsoonal patterns or droughts in other regions. This effort is hampered to date, however, because there are few scenarios for abrupt climate change that have been handed off by geoscientists to the economic modelers.

Attention to climate extremes in the wake of anthropogenic-induced gradual warming has been reported by Easterling et al. (2000). This study examined outputs of general circulation models that show changes in extreme events for future climates under greenhouse warming scenarios, such as increases in high temperatures, decreases in extreme low temperatures, and increases in precipitation events. These authors suggested a range of impacts due to these extremes, including the impacts to natural ecosystems and society.

In summary, climate change inevitably has impacts. Abruptness increases those impacts, especially on unmanaged and long-lived systems. To date, however, relatively little research has addressed the possible costs of abrupt climate change or ways to reduce these costs, both because climatologists have not produced appropriate scenarios and because ecological and economic scientists have not concentrated on abruptness. Ways to address this shortcoming are discussed in the next chapter.

6

Findings and Recommendations

he earth's climate system is characterized by change on all time and space scales, and some of the changes are abrupt even relative to the short time scales of relevance to human societies. Paleoclimatic records show that large, widespread, abrupt climate changes have affected much or all of the earth repeatedly over the last ice-age cycle as well as earlier – and these changes sometimes have occurred in periods as short as a few years. Perturbations in some regions were spectacularly large: some had temperature increases of up to 16°C and doubling of precipitation within decades, or even single years. Changes in precipitation and evaporation are estimated to have caused changes in the extent of wetlands around the world of up to 50 percent. Agreement between proxy and instrumental records and between different proxy records lends confidence to paleoclimatic reconstructions and allows scientists to be very confident that abrupt climate change is a real, recurrent phenomenon.

Abrupt climate changes in the last few thousand years generally have been less severe and affected smaller areas than some of the changes further back in the past. Nonetheless, evidence shows that rapid climate changes have affected societies and ecosystems substantially, especially when the changes that brought persistent droughts occurred in regions with human settlements.

There is no reason to believe that abrupt climate changes will not occur again. Furthermore, the paleoclimatic record demonstrates that the most

153

dramatic shifts in climate have occurred when factors controlling the climate system were changing. This has important implications for future climate in that it suggests that increasing human perturbation of the earth system may make abrupt change more likely.

Some sectors of the economy and ecosystems might be highly sensitive to abrupt climate change. Because the economy is more "managed" than are ecosystems, particularly in the industrial sectors of developed societies such as the United States, the major vulnerability to the effects of abrupt climate change is likely to lie at the intersection of human societies and ecosystems, such as for agriculture, forests, and water systems.

It is important not to be fatalistic about the threats posed by abrupt climate change. Societies have faced both gradual and abrupt climate changes for millennia and have learned to adapt through various mechanisms, such as moving indoors, developing irrigation for crops, and migrating away from inhospitable regions. Nevertheless, because climate change is likely to continue and may even accelerate in the coming decades, denying the likelihood or downplaying the relevance of past abrupt changes could be costly. Societies can take steps to face the potential for abrupt climate change. The committee believes that increased knowledge is the best way to improve the effectiveness of response, and thus that research into the causes, patterns, and likelihood of abrupt climate change can help reduce vulnerabilities and increase our adaptive capabilities. The committee's research recommendations fall into two broad categories: (1) implementation of targeted research to expand instrumental and paleoclimatic observations, and (2) implementation of modeling and associated analysis of abrupt climate change and its potential ecological, economic, and social impacts. What follows is a discussion of recommended research activities to support these two themes.

IMPROVE THE FUNDAMENTAL KNOWLEDGE BASE RELATED TO ABRUPT CLIMATE CHANGE

Recommendation 1. Research programs should be initiated to collect data to improve understanding of thresholds and nonlinearities in geophysical, ecological, and economic systems. Geophysical efforts should focus especially on modes of coupled atmosphere-ocean behavior, oceanic deepwater processes, hydrology, and ice. Economic and ecological research should focus on understanding nonmarket and environmental issues, initiation of a comprehensive land-use census, and development of integrated economic and ecological data sets. These data will enhance understanding of abrupt climate change impacts and will aid development of adaptation strategies.

A major finding of this study is that the only thing we can be sure of is that there will be climatic surprises. Physical, ecological, and human systems are complex, nonlinear, dynamic and imperfectly understood. Climate changes are producing conditions outside the range of recent historical experience and observation, and it is unclear how the systems will interact with and react to the coming climatic changes.

Data Needed to Better Understand the Mechanisms and Triggers of Abrupt Climate Change

It is crucial to be able to recognize present or impending abrupt climate changes quickly. This capability will involve improved monitoring of parameters that describe climatic, ecological, and economic systems. Some of the desired data are not uniquely associated with abrupt climate change and, indeed, have broad applications. Other data take on particular importance because they concern properties or regions implicated in postulated mechanisms of abrupt climate change, such as the strength of the Atlantic thermohaline circulation. Research to increase our understanding of abrupt climate change should be designed specifically within the context of the various mechanisms thought to be involved. Focus is required to provide data for process studies from key regions where triggers of abrupt climate change are likely to occur, and to obtain reliable time series of climate indicators that play crucial roles in the postulated mechanisms. Observations could enable early warning of the onset of abrupt climate change. New observational techniques and data-model comparisons (data assimilation) will be required as appropriate.

Circulation of the ocean is inferred primarily by combining measurements of water properties with physical understanding of flow. Direct measurements of currents and tracer distributions confirm the inferences. Data limitations, such as the technical difficulty of measuring deep-ocean currents directly, have left some uncertainties, but recent technical advances now allow much improved observations. Collection of more direct data on ocean circulation, together with targeted tracer studies, would reveal characteristics of ocean circulation, including the strength of deep water sources. The new knowledge will help in understanding active processes, such as deep convection and subduction, and perhaps could provide an early warning of changes in circulation that could lead to an abrupt climate change. If we are to develop predictive capabilities regarding the thermohaline circulation, we must observe its strength and structure. To date, however, no observational network exists to observe the thermohaline circulation on a continuous basis. Systematic, long-term observations of the heat and water fluxes influencing the thermohaline circulation are needed, especially in the areas of Northern Hemisphere deepwater formation. Moreover, remote influences on the thermohaline circulation must be monitored, particularly the low-latitude atmospheric water-vapor transport from the Atlantic to the Pacific and the influence of Southern Ocean changes.

The meaning of trends in sea-ice volume and extent and associated freshwater fluxes and how these will affect the potential for abrupt climate change is still debated. The debate is due, in part, to limitations in the data especially on ice thickness. Data collected from submarines provide some insight into changes in Arctic ice, but generation of long-term data is uncertain because of the planned cessation of submarine science cruises. Advancing the science of abrupt climate change requires improved observations of sea-ice extent, thickness, and fractional coverage.

The volume and extent of portions of the Greenland and Antarctic ice sheets are known to be changing rapidly, but full coverage by altimetry and interferometric synthetic-aperture radar used to measure thickness and velocity change of ice flow is not yet available, and near-polar holes in other remotely sensed fields miss key parts, especially in Antarctica. Basal conditions of the large ice sheets, and thus their potential for crossing thresholds and leading to rapid climate changes, are known at only a handful of points and with less confidence along limited aerogeophysical flight lines; most of the ice-sheet beds are uncharacterized. Processes beneath the floating extensions called ice shelves, where ice-sheet stability meets deepwater formation, are poorly known. In light of the clear paleoclimatic evidence of abrupt ice-sheet changes affecting global climate and sea level, enhanced emphasis on ice-sheet characterization over time is essential. In addition to the major ice sheets of Greenland and Antarctica, mountain glaciers also offer research opportunities: they are sensitive to climate change and can serve as sentinels of change, so improved monitoring and understanding of these systems is also needed.

Land hydrology links the atmosphere to oceans and ecosystems. Most terrestrial rainfall and snowmelt are used by plants, and much of the remainder runs off in rivers, which locally freshen seawater. Terrestrial precipitation is stored for highly variable periods in surface water, groundwater, glaciers, and ground ice and permafrost. Changes in the total flux of precipitation and in the balance among the components of the hydrological cycle can affect ocean circulation, vegetation types and productivity, and the availability of freshwater for human and ecosystem needs. Despite the obvious importance of the hydrological cycle, surprisingly little is known about total groundwater storage and water quality, trends in recharge and discharge, permafrost evolution (including feedbacks on greenhouse gases), feedbacks related to the generation and persistence of drought, how drought is eventually broken, and related topics. A concerted effort should be made to monitor essential components of the hydrological cycle, as this system is likely to be greatly affected by abrupt changes in climate.

Sophisticated systems are available to measure atmospheric properties and circulation, to assimilate new and old data, and thus to characterize the behavior of the atmosphere. Some key data sets, such as those from weather ships, remain vulnerable to funding cuts, and additional sampling in some regions would be useful (NRC, 1999d). Reliable assimilation of well-calibrated data sets is required to characterize gradual and abrupt climate change observed in instrumental records. Furthermore, additional analyses of the instrumental records are needed to identify climatic modes and their long-term behavior.

Data Needed to Understand the Effects of Climate Change on Ecological and Economic Systems

Understanding of how abrupt climate change will affect economic and ecological systems requires improved monitoring, detection, and measurements of critical components of nonmarket and ecological systems. Monitoring might provide early warning of climate-change effects and help minimize their influence. On the whole, the market economy in developed nations is well monitored in most sectors through national economic measures and accounts. Measures for nonmarket and ecological systems are much less systematic. Various research and monitoring activities should be considered:

• Efforts to identify plant and animal species should continue because this information is essential for determining extinction rates.
• High-frequency monitoring of terrestrial ecosystems is needed be-

cause this is critical for understanding the consequences of future climatic changes.

- Sustained monitoring of freshwater and marine ecosystems should be expanded.
- Alternative fire-monitoring and fire-management approaches should be studied, such as those involving improved coordination with seasonal climate forecasts, fuel data information, and tree thinning practices.
- Monitoring of wildlife diseases should be increased and collaboration between climate-change ecologists and the infectious-disease researchers should be encouraged.

It will also be important to link monitoring efforts in research institutions with databases at the National Wildlife Health Center of the US Geological Survey, the Animal and Plant Health Inspection Service, and the Centers for Disease Control and Prevention.

An increased understanding of the important thresholds and nonlinearities in economic and ecological systems will enable better prediction of effects of abrupt climate change. This report has emphasized that the most severe effects of abrupt climate change will occur when thresholds in climatic, ecological, and economic systems are crossed. A research priority is to develop a list of potential thresholds and feedbacks important to abrupt climate change, to quantify the location of the thresholds and strengths of feedbacks, and to improve the monitoring of systems related to these thresholds.

It is also important to develop measures of nonmarket economic sectors. Market accounts have been intensively studied and developed, but economic measures of nonmarket sectors have lagged far behind those of market sectors. A recent study by the National Research Council recommended the development of environmental and nonmarket accounts on a priority basis to improve management and understanding of nonmarket systems (Nordhaus and Kokkelenberg, 1999). Better measures of nonmarket sectors would provide necessary data for understanding the vulnerability of such systems to abrupt climate change.

One of the major concerns about the potential for ecosystem adaptation to climate change is the fragmentation of landscapes and ecosystems. A comprehensive land-use census is needed (NRC, 2000) to monitor the fragmentation of landscape and ecosystems and to provide information for helping reduce their vulnerability to abrupt climate change. There is no comprehensive census of land use in the United States and other countries. If a census were undertaken periodically, trends in fragmentation could be tracked, and potential steps to reduce the vulnerability to fragmentation could be better evaluated. The construction of databases to integrate economic, ecological, archeological, and climatic data would help to improve the understanding of the effects of past abrupt climate changes.

Some data collection required for better monitoring and understanding of the causes, magnitudes, and effects of abrupt climate change could be done in the private sector under a granting mechanism; other datasets might require federal data-collection efforts.

IMPROVE MODELING FOCUSED ON ABRUPT CLIMATE CHANGE

Recommendation 2. New modeling efforts that integrate geophysical, ecological, and social-science analyses should be developed to focus on investigating abrupt climate changes. In addition, new mechanisms that can cause abrupt climate change should be investigated, especially those operating during warm climatic intervals. Understanding of such mechanisms should be improved by developing and applying a hierarchy of models, from theory and conceptual models through models of intermediate complexity, to high-resolution models of components of the climate system, to fully coupled earth-system models. Model-data comparisons should be enhanced by improving the ability of models to simulate changes in quantities such as isotopic ratios that record past climatic conditions. Modeling should be used to generate scenarios of abrupt climate change with high spatial and temporal resolution for assessing impacts and testing possible adaptations. Enhanced, dedicated computational resources will be required for such modeling.

Developing theoretical and empirical models to understand abrupt climate changes and the interaction of such changes with ecological and economic systems has a high priority. Modeling is essential for collaborative research between physical, ecological, and social scientists, and much more effort in the development of accurate models will be necessary to produce a useful understanding of abrupt climate processes. Model analyses help to focus research on possible causes of abrupt climate change, such as human activities; on key areas where climatic thresholds might be crossed; and on fundamental uncertainties in climate-system dynamics. Most analyses have considered only gradual climate change; given the accumulating evidence of abrupt climate change and of its capacity to affect human societies, more attention should be focused on improving climate and impact-assessment models and on producing model scenarios involving abrupt climate change.

Climate models that are used to test leading hypotheses for abrupt climate change, such as altered deep-ocean circulation, can only partially simulate the size, speed, and extent of the large climatic changes that have occurred. The failure to explain the climate record fully suggests either that the proposed mechanisms being used to drive these models are incomplete or that the models are not as sensitive to abrupt climate change as is the natural environment. It is also of concern that existing models do not accurately simulate warm climates of the past.

Improved understanding of abrupt climatic changes that occurred in the past and are thus possible in the future can be gained through enhancements of climate models. A comprehensive modeling strategy designed to address abrupt climate change should include vigorous use of a hierarchy of models, from theory and conceptual models through models of intermediate complexity, to high-resolution models of components of the climate system, to fully coupled earth-system models. The simpler models are well suited for use in developing new hypotheses for abrupt climate change and should focus on warmer climates, because warming is likely. Because reorganizations of the thermohaline circulation have never been demonstrated in climate models employing high-resolution ocean components, improving the spatial resolution in climate models assumes high priority. Complex models should be used to produce geographically resolved (to about 1° of latitude by 1° of longitude), short-time (annual or seasonal) sensitivity experiments, and scenarios of possible abrupt climatic changes.

Long integrations of fully coupled models under various forcings for the past, present, and future will be required to evaluate the models, assess possibilities of future abrupt changes, and provide scenarios of those future changes. The scenarios can be combined with integrated-assessment economic models to improve understanding of the costs for alternative adaptive approaches to climate change with attention to the effects of rising greenhouse-gas concentrations and nonclimatic factors, such as land use changes and urbanization. Model-data comparisons are needed to assess the quality of model predictions. It is important to note that the multiple long integrations of enhanced, fully coupled earth-system models required for this research are not possible with the computer resources available today, and thus, these resources should be enhanced.

Ecologists and social scientists should use relevant scenarios of abrupt climate change for their modeling. An important reason why there have been so few studies on the impacts of future abrupt climate change is that climate scientists have provided social scientists little to work with; only the cooling events of the Little Ice Age and the Younger Dryas have been extensively analyzed, and the database on these remains inadequate. Complicating the issue is the uncertainty of how probabilities of abrupt climate change will be affected by increasing greenhouse-gas concentrations and by nonclimatic factors, such as trends in land and water use and urbanization.

Many economic models require high-resolution climatic and ecological data as inputs. A useful target of spatial resolution is 1° of latitude by 1° of longitude, annual or seasonal temporal resolution, and temperature and precipitation as a minimum for climatic variables. Up to now, scenarios at this resolution have been prepared only for gradual climate change; to be useful in developing impact analyses and projections, such high-resolution scenarios will be necessary for a variety of scenarios of abrupt climate change.

IMPROVE PALEOCLIMATIC DATA RELATED TO ABRUPT CLIMATE CHANGE

Recommendation 3. The quantity of paleoclimatic data on abrupt change and ecological responses should be enhanced, with special emphasis on:

• Selected coordinated projects to produce especially robust, multiparameter, high-resolution histories of climate change and ecological response.
• Better geographic coverage and higher temporal resolution.
• Additional proxies, including those that focus on water (e.g., droughts, floods, etc.).
• Multidisciplinary studies of selected abrupt climate changes.

Abrupt climate change is evident in model results and in instrumental records of the climate system, but much of today's interest in the subject was motivated by the strong evidence in paleoclimatic archives of extreme changes. Proxy records of paleoclimate are central to the subject and will continue to be so for some time.

Available paleoclimate records provide information on many environmental variables, such as temperature, moisture, and wind speed and direction. Temporal resolution ranges from subannual to multimillennial. Dating uncertainties are in a similar range. Many of the limitations on the climatic variables that can be reconstructed, their time resolution, and dating uncertainty are set by nature, but others are related to available resources. With more resources and effort, it will be possible to advance our understanding of abrupt climate change (NRC, 1999c).

Intensive, multiparameter, often multi-investigator projects can be especially valuable. As one example, in the ice-core projects from central Greenland, duplication of the measurements by independent international teams provides exceptional confidence in most data and reveals which datasets do not warrant confidence. Sampling at very high time resolution to produce datasets complementary to those of other investigators gives an exceptionally clear picture of past climate. Such projects require more funding and effort than are typical of paleoclimatic research, but they provide an essential reference standard, or "type section," of abrupt climate change to which other records can be compared. A difficulty is that this reference standard is from one place in high northern latitudes and is inappropriate for study of much of the climate system. Not all paleoclimatic records can be studied in the same detail as those from Greenland, but generation of at least a few similar highly resolved (preferably annually or subannually) reference standards, including a North Atlantic marine record comparable with Greenland records, would be of great value. The ultimate goal is to develop a global network of records with at least decadal resolution. Terrestrial and marine records of climate change and ecological response from the regions of the western Pacific warm pool (the warmest part of the global climate system) and the Southern Ocean and Antarctic continent (the southern cold pole of the climate system) are among the most critical targets for future paleoclimate research, including generation of reference standards.

Temperature is probably the easiest climatic variable to measure, and the public often focuses on it. But water availability over land is probably more important to economic and ecological systems. Focus on measures of precipitation, evaporation, and the quantitative difference between them is particularly important. For investigations of circulation involving the deep ocean, reconstructions of water-mass density in polar and subpolar regions are central. More broadly, improved efforts to develop and calibrate more proxies for additional paleoclimatic indicators are required to understand past changes fully.

Traditional "time-slice" reconstructions (e.g., the world 6,000 years ago) have provided insight into climate processes but are not the best vehicle for studying abrupt climate changes. Time-slices run the risk of mixing observations from just before and just after regional or larger abrupt changes. However, pending the still distant availability of high-time-resolution global maps of climate for all periods, anomaly-mapping efforts (e.g., perhaps DryMAP for the Younger Dryas, 8MAP for the event about 8,200 years ago, and LIMAP for the Little Ice Age) would be of considerable value. Similar studies of megadroughts are especially important. These efforts need to focus on the spatial and temporal variability of climatic changes and the resulting economic and ecological impacts. To place the warming and associated changes of the last hundred years in context and compare them with natural fluctuations, focus on the last 2,000 years is required.

IMPROVE STATISTICAL APPROACHES

Recommendation 4. Current practices in the development and use of statistics related to climate and climate-related variables generally assume a simple, unchanging distribution of outcomes. This assumption leads to serious underestimation of the likelihood of extreme events. The conceptual basis and the application of climatic statistics should be re-examined with an eye to providing realistic estimates of the likelihood of extreme events.

The term "climate" implies a relatively persistent set of environmental conditions. Because of this, some early reports of "abrupt climate change" were considered oxymorons, because if climate is a 30-year average, it cannot change in less than 30 years. This underscores the difficulty in recognizing, documenting, and discussing abrupt climate change. The difficulty is exacerbated by the observation of increased variability in some records near climate transitions and by the tendency of people to observe mode shifts in system behavior when only normal, uncorrelated random behavior is involved (Albright, 1993).

Many societal decisions are based on assumptions about the distributions of extreme weather-related events. Large capital projects—such as dams, airports, tunnels, subway systems, roads, levees, hydroelectric projects, and bridges—have embedded safety margins that are derived from data and assumptions about the frequency distribution of floods, hurricanes, storms, precipitation, and snowpacks. Many economic and business decisions depend on explicit or implicit assumptions about the distribution of weather-related events; examples of such decisions include the ratings of catastrophe bonds or weather derivatives; calculations that statisticians use to adjust data for seasonal effects; and public and private decisions made in agriculture, insurance, reinsurance, health (e.g., disease), fire control, and ecosystem management. Many of the decisions are based on statistical calculations that are appropriate for stationary climates. One well-known example is the widespread use of "30-year normals" in deriving climate data for individual locations.

On the whole, these assumptions are reasonable, if imperfect, rules of thumb to use when the variability of weather is small and climate is stationary. If some parameter related to climate, such as runoff, follows normal distributions with known and constant means and standard deviations, businesses and governments can reasonably use current practices. However, in light of recent findings related to nonstationary and often highly skewed climate-related variables, current practices can be misleading and result in costly errors. An event that is estimated to be a "500-year flood" or a "500-year fire" can be found to occur every 100 or 50 or 20 years if the averages and the variabilities are significantly misestimated.

The potential for abrupt climate change and the existence of thresholds for its effects require revisions of our statistical estimates and practices. We have not attempted to compile a list of all the places where society estimates or uses the assumption of a stationary and stable climate; such an effort would be enormous, and lies outside the committee charge. Rather, we believe that the time is ripe to examine this issue, develop improved methods, and develop improved estimates of the likelihood of extreme events.

INVESTIGATE "NO-REGRETS" STRATEGIES TO REDUCE VULNERABILITY

Recommendation 5. Research should be undertaken to identify "no-regrets" measures to reduce vulnerabilities and increase adaptive capacity at little or no cost. No-regrets measures may include low-cost steps to: slow climate change; improve climate forecasting; slow biodiversity loss; improve water, land, and air quality; and develop institutions that are more robust to major disruptions. Technological changes may increase the adaptability and resiliency of market and ecological systems faced by the prospect of damaging abrupt climate change. Research is particularly needed to assist poor countries, which lack both scientific resources and economic infrastructure to reduce their vulnerabilities to potential abrupt climate changes.

This report emphasizes that social and ecological systems have long dealt with climate variability. Societies have taken many steps over the years, decades, and millennia to reduce their vulnerability to the effects of climate change, and ecosystems have weathered ice ages and extreme climatic events, such as the Younger Dryas. To some extent, it might be possible to reduce vulnerability and increase adaptation at little or no cost, by nudging research and policy in directions that will increase the adaptability of systems (NRC, 1992). Put differently, some current policies and practices may be ill advised and may prove inadequate in a world of rapid and unforeseen climatic changes. Improving such inadequate policies would be beneficial even if abrupt climate change turns out to fit a best-case, rather than a worst-case, scenario. Societies would have "no regrets" about the new policies, because they will be good policies in fair as well as in foul weather.

This report cannot provide a complete catalog of potential no-regrets options, but a few areas of interest are highlighted below. Research in these areas may lead to useful policy recommendations.

• *Energy policies*. Earlier National Research Council reports have identified policies that would slow climate change with low or even negative costs. For example, the phaseout of chlorofluorocarbons over the last 2 decades and replacement with gases with typically shorter atmospheric lifetimes has reduced the US contribution to global warming while also reducing future health risks posed by ozone depletion (NRC, 1992). Furthermore, moving away from coal-burning toward other fuels, particularly natural gas, would reduce greenhouse-gas emissions and some effects on health and the environment, and might prove beneficial in the long term.

• *Ecological policies*. In land-use and coastal planning, managers may be helped by information on the effects on ecosystem services of nonlinear future changes in climate. Scientists and government organizations at various levels could collaborate to develop and implement regulations and policies that reduce environmental degradation of water, air, and biota. Conservation measures related to land and watersheds may reduce the rate of biotic invasions, and management strategies may limit the spread of invasions. The economic and ecological costs of disease emerging from abrupt climate change may help guide response.

• *Forecasting of weather and weather-related events*. Hurricanes and other storms can have large impacts. Climate scientists are uncertain how climate change will affect the frequency and intensity of severe storms, but changes are possible, and societies can reduce vulnerabilities through improved forecasting. It is striking that, although normalized property damages caused by hurricanes in the United States in the twentieth century did not decline (because most houses cannot move when hurricanes are forecast), the loss of life was markedly reduced because better forecasting allowed people to evacuate.

• *Institutions.* One of the most promising options is to conduct research leading to the development of improved institutions that will allow societies to withstand the greater risks that could be associated with abrupt climate change. Three examples—water systems, insurance, and statistical data—highlight the possibilities:

Water systems. Most evidence points to the likelihood that water systems will be severely stressed in the coming decades, and marginal regions face the possibility of droughts. Research can help identify better ways to manage water, such as the use of water markets or other flexible or innovative approaches that might prove to be useful strategies under scenarios of abrupt climate change (Intergovernmental Panel on Climate Change, 2001a).

Insurance. An important way that individuals or even countries can adapt to extreme events is through insurance against the ruinous effects of fires, floods, storms, and hurricanes. In essence, the existence of insurance promotes risk-taking behavior. Because losses from some of these risks are highly skewed, insurance payouts from time to time will be extremely large and will threaten to exhaust the reserves of insurance companies. Through the development of new instruments, such as weather derivatives and catastrophe bonds, markets can accommodate extreme events. One important adaptation would be the development of better instruments to spread the large losses ($50 billion and up) attributed to extreme weather and other events; they should be priced realistically to reflect the risks to discourage excessive risk-taking. Research into the value of such better instruments is needed.

Statistical data. Analysis of statistical data may help identify beneficial, low-cost ways to reform rules concerning flood-plain and coastal management.

FINAL THOUGHTS

Because of the strength of existing infrastructure and institutions, the United States and other wealthy nations are likely to cope with the effects of abrupt climate change more easily than poorer countries. This does not mean that developed countries can remain isolated from the rest of the world, however. With growing globalization, adverse impacts—although likely to vary from region to region because exposure and sensitivity will vary—are likely to spill across national boundaries, through human and biotic migration, economic shocks, and political aftershocks. Thus, even though this report focuses primarily on the United States, the issues are global and it will be important to give attention to the issues faced by poorer countries that are likely to be especially vulnerable to the social and economic impacts of abrupt climate change.

The United States is uniquely positioned to provide both scientific and financial leadership, and to work collaboratively with scientists around the world, to gain better understanding of the global impacts of abrupt climate change as well as reducing the vulnerability and increasing the adaptation in countries that are particularly vulnerable to these changes. Many of the recommendations in this report, although currently aimed at US institutions, would apply throughout the world.

References

Aagard, K., and E.C. Carmack. 1989. The role of sea ice and other fresh water in the Arctic circulation. *Journal of Geophysical Research* 94:14,485-14,498.

Abramovitz, J.N. 1996. Imperiled waters, impoverished future: The decline of freshwater ecosystems. Peterson, editor. Worldwatch Institute, Washington, D.C.

Aeberhardt, M., M. Blatter, and T.F. Stocker. 2000. Variability on the century time scale and regime changes in a stochastically forced zonally averaged ocean-atmosphere model. *Geophysical Research Letters* 27:1303-1306.

Ágústsdóttir, A.M., R.B. Alley, D. Pollard, and W. Peterson. 1999. Ekman transport and upwelling from wind stress from GENESIS climate model experiments with variable North Atlantic heat convergence. *Geophysical Research Letters* 26(9):1333-1336.

Ahmad, J., G. Beg Paklar, F. Bonjean, G. Dal Bo', T. Kaminski, F. Kauker, M. Monai, E. Pierazzo, P. Stone, and L. Zampato. 1997. Pp. 87-100 in *The interaction between global warming and the thermohaline circulation*. Environmental Dynamics Series, Vol. 4, Istituto Veneto di Scienze, Lettere ed Arti.

Albright, S.C. 1993. A statistical-analysis of hitting streaks in baseball. *Journal of the American Statistical Association* 88:1175-1183.

Allen, C. 2001. *Rapid, Climate-Induced Ecological Disturbances: Forest Dieback, Erosion, and Fire*. Summary notes from Abrupt Climate Change Meeting, National Academy of Sciences, March 21-22, Washington, D.C.

Allen, C.D., and D. Breshears. 1998. Drought-induced shift of a forest-woodland ecotone: rapid landscape response to climate variation. *Proceedings of the National Academy of Sciences of the United States of America* 95:14,839-14,842.

Alley, R.B. 2000. The Younger Dryas cold interval as viewed from central Greenland. *Quaternary Science Reviews* 19:213-226.

Alley, R.B., and D.R. MacAyeal. 1994. Ice-rafted debris associated with binge/purge oscillations of the Laurentide Ice Sheet. *Paleoceanography* 9:503-511.

Alley, R.B., and P.U. Clark. 1999. The deglaciation of the Northern Hemisphere: A global perspective. *Annual Reviews of Earth and Planetary Sciences* 27:149-182.

Alley, R.B., and R.A. Bindschadler. 2000. The West Antarctic Ice Sheet and sea-level change. Pp. 1-11 in R.B. Alley and R.A. Bindschadler (eds.), *The West Antarctic Ice Sheet: Behavior and Environment.* Antarctic Research Series, Vol. 77. American Geophysical Union, Washington, D.C.

Alley, R.B., D.A. Meese, C.A. Shuman, A.J. Gow, K.C. Taylor, P.M. Grootes, J.W.C. White, M. Ram, E.D. Waddington, P.A. Mayewski, and G.A. Zielinski. 1993. Abrupt increase in Greenland snow accumulation at the end of the Younger Dryas event. *Nature* 362:527-529.

Alley, R.B., R.C. Finkel, K. Nishiizumi, S. Anandakrishnan, C.A. Shuman, G.R. Mershon, G.A. Zielinski, and P.A. Mayewski. 1995. Changes in continental and sea-salt atmospheric loadings in central Greenland during the most recent deglaciation. *Journal of Glaciology* 41(139):503-514.

Alley, R.B., A.J. Gow, S.J. Johnsen, J. Kipfstuhl, D.A. Meese, and T. Thorsteinsson. 1995a. Comparison of deep ice cores. *Nature* 373:393-394.

Alley, R.B., R.C. Finkel, K. Nishiizumi, S. Anandakrishnan, C.A. Shuman, G.R. Mershon, G.A. Zielinski, and P.A. Mayewski. 1995b. Changes in continental and sea-salt atmospheric loadings in central Greenland during the most recent deglaciation. *Journal of Glaciology* 41(139):503-514.

Alley, R.B., P.A. Mayewski, T. Sowers, M. Stuiver, K.C. Taylor, and P.U. Clark. 1997. Holocene climatic instability: A prominent, widespread event 8,200 yrs. ago. *Geology* 25(6):483-486.

Alley, R.B., A.M. Ágústsdóttir, and P.J. Fawcett. 1999. Ice-core evidence of Late-Holocene reduction in north Atlantic ocean heat transport. Pp. 301-312 in P.U. Clark, R.S. Webb and L.D. Keigwin (eds.), *Mechanisms of Global Climate Change at Millennial Time Scales,* Geophysical Monograph 112. American Geophysical Union, Washington, D.C.

Alley, R.B., S. Anandakrishnan, and P. Jung. 2001. Stochastic resonance in the North Atlantic. *Paleoceanography,* 16(2):190-198.

Andrewartha, H.G., and L.C. Birch. 1954. *The Distribution and Abundance of Animals.* University of Chicago Press, IL.

Andrews, J.T., and K. Tedesco. 1992. Detrital carbonate-rich sediments, northwestern Labrador Sea: Implications for ice-sheet dynamics and iceberg rafting (Heinrich) events in the North Atlantic. *Geology* 20:1087-90.

Arnell, N.W. 2000. Thresholds and response to climate change forcing: The water sector. *Climatic Change* 46(3):305-316.

Bains, S., R.M. Corfield, and R.D. Norris. 1999. Mechanisms of climate warming at the end of the Paleocene. *Science* 285:724-727.

Baker, V. 1998. Paleohydrology and the hydrological sciences. In G. Benito, V. Baker, and K. Gregory (eds.), *Palaeohydrology and Environmental Change.* John Wiley and Sons, Ltd., New York.

Baker, V. 2000. Paleoflood hydrology and the estimation of extreme floods. In Wohl, E. (ed.), *Inland Flood Hazards.* Cambridge University Press, UK.

Balbus, J.M., and M.L. Wilson. 2000. *Human Health and Global Climate Change.* Report of the Pew Center on Global Climate Change. Pew Center on Global Climate Change, Arlington, Virginia. 43 pp.

Baldwin, M.P., and T.J. Dunkerton. 1999. Downward propagation of the Arctic Oscillation from the stratosphere to the troposphere. *Journal of Geophysical Research* 104:30,937-30,946.

Balling, R.C., Jr. 1988. The climatic impact of Sonoran vegetation discontinuity. *Climate Change* 13:99-109.

Barber D.C., A. Dyke, C. Hillaire-Marcel, A.E. Jennings, J.T. Andrews, M.W. Kerwin, G. Bilodeau, R. McNeely, J. Southon, M.D. Morehead, and J.M. Gagnon. 1999. Forcing of the cold event of 8,200 years ago by catastrophic drainage of Laurentide lakes. *Nature* 400 (6742):344-348.

Bard, E., R. Rostek, and C. Sonzogni. 1997. Interhemispheric synchrony of the last deglaciation inferred from alkenone palaeothermometry. *Nature* 385:707-710.

Barnett, T.P., D.W. Pierce, and R. Schur. 2001. Detection of climate change in the world's oceans. *Science* 292:270-274.

Barron, E.J. 1987. Eocene equator-to-pole surface temperatures: A significant climate problem. *Paleoceanography* 2:729-739.

Beaulieu, J.L., V. de Andrieu, P. Ponel, M. Reille, and J.J. Lowe. 1994. The Weichselian lateglacial in southwestern Europe (Iberian Peninsula, Pyrenees, Massif Central, northern Apennines). *Journal of Quaternary Science* 9:101-107.

Behl, R.J., and J.P. Kennett. 1996. Brief interstadial events in the Santa Barbara basin, NE Pacific, during the past 60 kyr. *Nature* 379:243-246.

Belkin, I.M., S. Levitus, J. Antonov, and S.A. Malmberg. 1998. Great salinity anomalies in the North Atlantic. *Progress in Oceanography* 41:1-68.

Bender, M., T. Sowers, M.L. Dickson, J. Orchardo, P. Grootes, P.A. Mayewski, and D.A. Meese. 1994. Climate correlations between Greenland and Antarctica during the past 100,000 years. *Nature* 372:663-666.

Bender, M., B. Malaize, J. Orchardo, T. Sowers, and J. Jouzel, 1999. High precision correlations of Greenland and Antarctic ice core records over the last 100 kyr. Pp. 149-164 in P.U. Clark, R.S. Webb, and L.D. Keigwin (eds.), *Mechanisms of Global Climate Change at Millennial Time Scales*. Geophysical Monograph 112. American Geophysical Union, Washington, D.C.

Ben-Gai, T., A. Bitan, A. Manes, P. Alpert, and K. Israeli. 1998. Aircraft measurements of surface albedo in relation to climatic changes in southern Israel. *Theoretical Applications in Climatology* 61:207-215.

Benito, G., V.R. Baker, and K.J. Gregory. 1998. *Palaeohydrology and Environmental Change*. Wiley, New York. 353pp.

Berger, L., R. Speare, P. Daszak, D. E. Green, A.A. Cunningham, C.L. Goggin, R. Slocombe, M.A. Ragan, A.D. Hyatt, K.R. McDonald, H.B. Hines, K.R. Lips, G. Marantelli, and H. Parkes. 1998. Chytridiomycosis causes amphibian mortality associated with population declines in the rainforests of Australia and Central America. *Proceedings of the National Academy of Sciences of the United States of America* 95(15): 9031-9036.

Berner, J., and G. Branstator. 2001. Consequences of nonlinearities on the low-frequency behaviour of an AGCM. Proceeding of the 13th Conference on Atmospheric and Oceanic Fluid Dynamics. *Bulletin of the American Meteorological Society*, July 2001, Preprints, pp. 231-234.

Bice, K.L., and J. Marotzke. In press. Could changing ocean circulation have destabilized methane hydrate at the Palaeocene/Eocene boundary? *Paleooceanography*.

Billings, W.D. 1987. Carbon balance of Alaskan tundra and taiga ecosystems: Past, present, and future. *Quaternary Science Reviews* 6:165-177.

Binford, M.W., A.L. Kolata, M. Brenner, J. Janusek, M.T. Seddon, M.B. Abbott, and J.H. Curtis. 1997. Climate variation and the rise and fall of an Andean civilization. *Quaternary Research* 47:235-248.

Birks, H.H., and B. Ammann. 2000. Two terrestrial records of rapid climatic change during the glacial-Holocene transition (14,000-9,000 calender years B.P) from Europe. *Proceedings of the National Academy of Sciences of the United States of America* 97(4):1390-1394.

Biscaye, P.E., F.E. Grousset, M. Revel, S. Van der Gaast, G.A. Zielinski, A. Vaars, and G. Kukla. 1997. Asian provenance of glacial dust (stage 2) in the Greenland Ice Sheet Project 2 Ice Core, Summit, Greenland. *Journal of Geophysical Research* 102:26,765-26,781.

Bjorck, S., N. Noe-Nygaard, J. Wolin, M. Houmark-Nielsen, H.J. Hansen, and I. Snowball. 2000. Eemian lake development, hydrology and climate: A multi-stratigraphic study of the Hollerup site in Denmark. *Quaternary Science Reviews* 19(6):509-536.

Bjorck, S., S. Olsson, C. EllisEvans, H. Hakansson, O. Humlum, and J.M. deLirio. 1996. Late Holocene palaeoclimatic records from lake sediments on James Ross Island, Antarctica. *Palaeogeography, Palaeoclimatology, Palaeoecology* 121(3-4):195-220.

Black, D.E., L.C. Peterson, J.T. Overpeck, A. Kaplan, M. Evans, and M. Kashgarian. 1999. Eight centuries of North Atlantic atmosphere-ocean variability. *Science* 286:1709-1713.

Blunier, T., and E.J. Brook. 2001. Timing of millennial-scale climate change in Antarctica and Greenland during the last glacial period. *Science* 291:109-112.

Blunier, T., J. Chappellaz, J. Schwander, B. Stauffer, and D. Raynaud. 1995. Variations in atmospheric methane concentration during the Holocene epoch. *Nature* 374:46-49.

Boag, P.T. and P.R. Grant. 1984. Darwin finches (Geospiza) on Isla Daphne Major, Galapagos —breeding and feeding ecology in a climatically variable environment. *Ecological Monographs* 54(4):463-489.

Boettger, T., F.W. Junge, and T. Litt. 2000. Stable climatic conditions in central Germany during the last interglacial. *Journal of Quaternary Science* 15(5):469-473.

Bond, G., W. Broecker, S. Johnsen, J. McManus, L. Labeyrie, J. Jouzel, and G. Bonani. 1993. Correlations between climate records for North Atlantic sediments and Greenland ice. *Nature* 365:143-147.

Bond, G., and R. Lotti. 1995. Iceberg discharges into the North Atlantic on millennial time scales during the last deglaciation. *Science* 267:1005-1010.

Bond, G., W. Showers, M. Chesby, R. Lotti, I. Hajdas, and G. Bonani. 1997. A pervasive millennial-scale cycle in North Atlantic Holocene and glacial climates. *Science* 278:1257-1266.

Bonnefille, R., G. Riollet, G. Buchet, M. Arnold, M. Icole, and R. Lafont. 1995. Glacial/interglacial record from intertropical Africa: High resolution pollen and carbon data at Rusaka, Burundi. *Quaternary Science Reviews* 14(9):917-936

Borchert, J.R. 1950. The climate of the Central North American grasslands. *Annual Association of American Geography* 40:1-39.

Bouma, M.J., and C. Dye. 1997. Cycles of malaria associated with El Niño in Venezuela. *Journal of the American Medical Association* 278:1772.

Bounoua, L., G.J. Collatz, C.J. Tucker, P.J. Sellers, D.A. Randall, D.A. Dazlich, T.G. Jensen, S.O. Los, J.A. Berry, C.B. Field, and I. Fung. 1999. Interactions between vegetation and climate: Radiative and physiological effects of doubled CO_2. *Journal of Climate* 12:309-324.

Boyle, E.A. 2000. Is ocean thermohaline circulation linked to abrupt stadial/interstadial transitions? *Quaternary Science Reviews* 19:255-272.

Boyle, E.A., and L.D. Keigwin. 1987. North Atlantic thermohaline circulation during the last 20,000 years linked to high latitude surface temperatures. *Nature* 330:35-40.

Braconnot, P., S. Joussaume, O. Marti, and N. de Noblet. 1999. Synergistic feedbacks from ocean and vegetation on the African monsoon response to mid-Holocene insolation. *Geophysical Research Letters* 26:2481-2484.

Bradley, R.S. 1999. *Paleoclimatology: Reconstructing Climates of the Quaternary*, 2nd edition. Academic Press, San Diego, CA. 613 pp.

Braudel, F. 1973. *Capitalism and Material Life, 1400-1800*. HarperCollins, New York, 18 pp.

Bravar, L., and M.L. Kavvas. 1991. On the physics of drought. I. A conceptual framework. *Journal of Hydrology* 129:281-297.

Brenner, M., D. Hodell, J. Curtis, M. Rosenmeier, M. Binford, and M. Abbott. 2001. Abrupt climate change and pre-Columbian cultural collapse. Pp. 87-101 in V. Markgraf (ed.), *Interhemispheric Climate Linkages*. Academic Press, New York.

Bretherton, C.S., and D.S. Battisti. 2000. An interpretation of the results from atmospheric general circulation models forced by the time history of the observed sea surface temperature distribution. *Geophysical Research Letters* 27:767-770.

Briffa, K.R., T.J. Osborn, F.H. Schweingruber, I.C. Harris, P.D. Jones, S.G. Shiyatov, and E.A. Vaganov. 2001. Low-frequency temperature variations from a northern tree ring density network. *Journal of Geophysical Research-Atmospheres* 106:2929-2941.

Broccoli, A.J., and S. Manabe. 1987. The influence of continental ice, atmospheric CO_2, and land albedo on the climate of the last glacial maximum. *Climate Dynamics* 1:87-99.

Broecker, W.S., and T.-H.Peng. 1982. *Tracers in the Sea*. Eldigio Press, Lamont-Doherty Earth Observatory, Palisades, NY.

Broecker, W.S. 1987. Commentary: Unpleasant surprises in the greenhouse? *Nature* 328: 123:126.

Broecker, W.S., M. Andree, W. Wolfli, H. Oeschger, G. Bonani, J. Kennett, and D. Peteet. 1988. The chronology of the last deglaciation: Implications to the cause of the Younger Dryas event. *Paleoceanography* 3:1-19.

Broecker, W.S. 1994. Massive iceberg discharges as triggers for global climate change. *Nature* 372:421-424.

Broecker, W.S. 1995. *The Glacial World According to Wally*, 2nd edition. Eldigio Press, Lamont-Doherty Earth Observatory of Columbia University, Palisades, NY.

Broecker, W.S. 1997. Thermohaline circulation, the Achilles heel of our climate system: Will man-made CO_2 upset the current balance? *Science* 278:1582-1588.

Broecker, W.S. 1998. Paleocean circulation during the last deglaciation: A bipolar seesaw? *Paleoceanography* 13:119-121.

Broecker, W.S. 1999. Was a change in the thermohaline circulation responsible for the Little Ice Age? *Proceedings of the National Academy of Sciences of the United States of America* 97:1339-1342.

Broecker, W.S. 2001. Was the Medieval Warm Period global? *Science* 291:1497-1499.

Broecker, W.S., and G.H. Denton. 1989. The role of ocean-atmosphere reorganization in glacial cycles. *Geochimica et Cosmochimica Acta* 53:2465-2501.

Broecker, W.S., D.M. Peteet, I. Hajdas, J. Lin, and E. Clark. 1998. Antiphasing between rainfall in Africa's Rift Valley and North America's Great Basin. *Quaternary Research* 50:12-20.

Broecker, W.S., D.M. Peteet, and D. Rind. 1985. Does the ocean-atmosphere system have more than one mode of operation? *Nature* 315:21-25.

Brook, E.J., T. Sowers, and J. Orchardo. 1996. Rapid variations in atmospheric methane concentration during the past 110,000 years. *Science* 273:1087-1093.

Brook, E.J., S. Harder, J. Severinghaus, and M. Bender. 1999. Atmospheric methane and millennial-scale climate change. *Geophysical Monograph* 112:165-176.

Brown, P., J.P. Kennett, and B.L. Ingram. 1999. Marine evidence for episodic Holocene megafloods in North America and the northern Gulf of Mexico. *Paleoceanography* 14(4):498-510.

Brown, L.R., C. Flavin, H. French, J.N. Abramowitz, C. Bright, S. Dunn, B. Halweil, G. Gardner, A. Mattoon, A. Platt McGinn, M. O'Meara, S. Postel, M. Renner, and L. Starke. 2000. *State of the World 2000*. W.W. Norton and Company, New York. 276 pp.

Brubaker, K., D. Entekhabi, and P. Eagleson. 1993. Estimation of continental precipitation recycling. *Journal of Climate* 6:1077-1089.

Bryan, F. 1986. High-latitude salinity effects and interhemispheric thermohaline circulations. *Nature* 323:301-304.

Bryan, F. 1987. On the parameter sensitivity of primitive equation ocean general circulation models. *Journal of Physical Oceanography* 17:970-985.

Bryant, N.A., L.F. Johnson, A.J. Brazel, R.C. Balling, C.F. Hutchinson, and L.R. Beck. 1990. Measuring the effect of over-grazing in the Sonoran Desert. *Climate Change* 17:243-264.

Bryden, H.L., D.H. Roemmich, and J.A. Church. 1991. Ocean heat transport across 24°N in the Pacific. *Deep-Sea Research* 38:297-324.

Bureau of Economic Analysis, 2001. http://www.bea.doc.gov/bea/dn1.htm.

Burroughs, W.J. 1999. *The Climate Revealed*. Cambridge University Press, UK. 192 pp.

Bush, A.B.G. 1999. Assessing the impact of mid-Holocene insolation on the atmosphere-ocean system. *Geophysical Research Letters* 26(1):99-102.

Bush, M.B., D.R. Piperno, P.A. Colinvaux, P.E. De Oliveira, L.A. Krissek, M.C. Miller, and W.E. Rowe. 1992. A 14,300-year paleoecological profile of a lowland tropical lake in Panama. *Ecological Monographs* 62:251-275.

Caldeira, K., and J.F. Kasting. 1992. Susceptibility of the early Earth to irreversible glaciation caused by carbon dioxide clouds. *Nature* 359:226-228.

Carmack, E.C., R.W. Macdonald, R.G. Perkin, and F.A. McLaughlin. 1995. Evidence for warming of Atlantic water in the southern Canada Basin. *Geophysical Research Letters* 22:1061-1064.

Carrington, D.P., R.G. Gallimore, and J.E. Kutzbach. 2001. Climate sensitivity to wetlands and wetland vegetation in mid-Holocene North Africa. *Climate Dynamics* 17(2-3):151-157.

Cavalieri D.J., P. Gloersen, C.L. Parkinson, J.C. Comiso, and H.J. Zwally. 2000. Observed hemispheric asymmetry in global sea ice changes. *Science* 278:1104-1106.

Chang, F.-C., and J.M. Wallace. 1987. Meteorological conditions during heat waves and droughts in the United States Great Plains. *Monthly Weather Review* 115:1253-1269.

Change, F.-C., and E.A. Smith. 2001. Hydrological and dynamical characteristics of summer-time drought over U.S. Great Plains. *Journal of Climate* 14:2296-2316.

Chapin III, F.S., E.S. Zavaleta, V.T. Eviner, R.L. Naylor, P.M. Vitousek, H.L. Reynolds, D.U. Hooper, S. Lavorel, O.E. Sala, S.E. Hobbie, M. Mack, and S. Diaz. 2000. Consequences of changing biodiversity. *Nature* 405:234-240.

Chapman, W.L., and J.E. Walsh. 1993. Recent variations of sea-ice and air temperature in high latitudes. *Bulletin of the American Meteorological Society* 74:33-47.

Chappellaz, J., T. Blunier, S. Kints, A. Dallenbach, J.M. Barnola, J. Schwander, D. Raynaud, and B. Stauffer. 1997. Changes in the atmospheric CH_4 gradient between Greenland and Antarctica during the Holocene. *Journal of Geophysical Research* 102:15,987-15,997.

Chappellaz, J., E. Brook, T. Blunier, and B. Malaize. 1997. CH_4 and $\delta^{18}O$ of O_2 records from Antarctic and Greenland ice: A clue for stratigraphic disturbance in the bottom part of the Greenland Ice Core Project and the Greenland Ice Sheet Project 2 ice cores. *Journal of Geophysical Research*. 102(C12):26,547-26,557.

Cheddadi, R., K. Mamakowa, J. Guiot, J.L. de Beaulieu, M. Reille, V. Andrieu, W. Granoszewski, and O. Peyron. 1998. Was the climate of the Eemian stable? A quantitative climate reconstruction from seven European pollen records. *Palaeogeography, Palaeoclimatology, Palaeoecology* 143(1-3):73-85.

Cheng, W. 2000. Climate variability in the North Atlantic: Interannual to multi-decadal time scales. Ph.D. Dissertation. Rosenstiel School of Marine and Atmospheric Sciences, University of Miami, FL.

Clark, D.B., Y. Xue, R.J. Harding, and P.J. Valdes. 2001. Modeling the impact of land surface degradation on the climate of tropical North Africa. *Journal of Climate* 14:1809-1822.

Claussen, M., C. Kubatzki, V. Brovkin, and A. Ganopolski. 1999. Simulation of an abrupt change in Saharan vegetation in the mid-Holocene. *Geophysical Research Letters* 26:2037-2040.

Clement A.C., Seager R., Cane M.A., 1999. Orbital controls on the El Niño/Southern Oscillation and the tropical climate. *Paleoceanography* 14(4):441-456.

Clement, A.C., R. Seager, and M.A. Cane. 2000. Supression of El Niño during the mid-Holocene by changes in the Earth's orbit. *Paleoceanography* 15:731-737.

Clement, A.C., M.A. Cane, and R. Seager. 2001. An orbitally driven tropical source for abrupt climate change. *Journal of Climate* 14:2369-2375.

Cole, J.E. 2001. A slow dance for El Niño. *Science* 291:1496-1497.

Cole, J.E., and E.R. Cook. 1998. The changing relationship between ENSO variability and moisture balance in the continental United States. *Geophysical Research Letters* 25:4529-4532.

Cole, J.E., J.T. Overpeck, and E.R. Cook. Submitted. Multiyear La Niña events and persistent drought in the contiguous United States. *Geophysical Research Letters*.

Colwell, R.R. 1996. Global climate and infectious disease: The cholera paradigm. *Science* 274(5295):2025-2031.

Conway, H., B.L. Hall, G.H. Denton, A.M. Gades, and E.D. Waddington. 1999. Past and future grounding-line retreat of the West Antarctic ice sheet. *Science* 286:280-283.

Cook, E.R., and A.H. Johnson. 1989. Climate change and forest decline: A review of the red spruce case. *Water, Air, and Soil Pollution* 48:127-140.

Corti, S., F. Molteni, and T.N. Palmer. 1999. Signature of recent climate change in frequencies of natural atmospheric circulation regimes. *Nature* 398:799-802.

Cortijo, E., L. Labeyrie, M. Elliot, E. Balbon, and N. Tisnerat. 2000. Rapid climate variability of the North Atlantic Ocean and global climate: A focus of the IMAGES program. *Quaternary Science Reviews* 19(1-5):227-241.

Crick, H.Q.P., C. Dudley, D. Glue, and D. Thomson. 1997. UK birds are laying eggs earlier. *Nature* 388:526.

Crick, H.Q.P., and T.H. Sparks. 1999. Climate change related to egg-laying trends. *Nature* 399:423.

Cronin T, D. Willard, A. Karlsen, S. Ishman, S. Verardo, J. McGeehin, R. Kerhin, C. Holmes, S. Colman, and A. Zimmerman. 2000. Climate variability in the eastern United States over the past millennium from Chesapeake Bay sediments. *Geology* 28:3-6.

Cronin, T.M. 1999. *Principles of Paleoclimatology*. Columbia University Press, New York.

Crowley, T.J. 1992. North Atlantic deep water cools the southern hemisphere. *Paleoceanography* 7:489-497.

Crowley, T.J. 2000. Causes of climate change over the past 1000 years. *Science* 289:270-277.

Crowley, T.J., and T.S. Lowery, 2000. How warm was the medieval warm period? *Ambio* 29:51-54.

Cuffey, K.M., R.B. Alley, P.M. Grootes, J.F. Bolzan, and S. Anandakrishnan. 1994. Calibration of the $\delta^{8}O$ isotopic paleothermometer for central Greenland, using borehole temperatures. *Journal of Glaciology* 40:341-349.

Cuffey, K.M., G.D. Clow, R.B. Alley, M. Stuiver, E.D. Waddington, and R.W. Saltus. 1995. Large Arctic temperature change at the glacial-Holocene transition. *Science* 270:455-458.

Cuffey, K.M., and G.D. Clow. 1997. Temperature, accumulation, and ice sheet elevation in central Greenland through the last deglacial transition. *Journal of Geophysical Research* 102(26):383-396.

Cuffey, K.M. and S.J. Marshall. 2000. Substantial contribution to sea-level rise during the last interglacial from the Greenland ice sheet. *Nature* 404(6778):591-594.

Cullen, H.M., P.B. deMenocal, S. Hemming, G. Hemming, F.H. Brown, T. Guilderson, and F. Sirocko. 2000. Climate change and the collapse of the Akkadian empire: Evidence from the deep sea. *Geology* 28:379-382.

Cwynar, L.C., and R.W. Spear. 2001. Late glacial climate change in the White Mountains of New Hampshire. *Quaternary Science Reviews* 20(11):1265-1274.

Dai, A., K.E. Trenberth, and T.R. Karl. 1999. Effects of clouds, soil moisture, precipitation and water vapor on diurnal temperature range. *Journal of Climate* 12:2451-2473.

Dällenbach, A., T. Blunier, J. Flückiger, B. Stauffer, J. Chappellaz, and D. Raynaud. 2000. Changes in the atmospheric CH_4 gradient between Greenland and Antarctica during the Last Glacial and the transition to the Holocene. *Geophysical Research Letters* 27:1005-1008.

Dansgaard, W., S.J. Johnsen, H.B. Clausen, D. Dahl-Jensen, N. Gundestrup, C.U. Hammer, and H. Oeschger. 1984. North Atlantic climatic oscillations revealed by deep Greenland ice cores. Pp. 288-298 in J. Hansen and T. Takahashi (eds.), *Climate Processes and Climate Sensitivity*. American Geophysical Union, Washington, D.C.

Daszak, P. 2001. University of Georgia. Personal communication. Social Impacts workshop.

Daszak, P., A.A. Cunningham, and A.D. Hyatt. 2001. Anthropogenic environmental change and the emergence of infectious diseases in wildlife. *Acta Tropica* 78(2):103-116.

Daszak, P., A.A. Cunningham, and A.D. Hyatt. 2000. Conservation conundrum - Response. *Science* 288:2320-2320.

Davis, M.B. 1983. Holocene vegetational history of the eastern United States. Pp. 166-181 in H.E. Wright, Jr. (ed.), *Late Quaternary Environments of the United States*, vol. 2. University of Minnesota Press, Minneapolis.

de la Mare, W.K. 1997. Abrupt mid-twentieth-century decline in Antarctic sea-ice extent from whaling records. *Nature* 389:57-60.

De Vries, J. 1976. *The Economy of Europe in an Age of Crisis*. Cambridge University Press, UK.

De Vries, J. 1981. Measuring the impact of climate on history: The search for appropriate methodologies. In R.I. Rotberg and T.K. Rabb (eds.), *Climate and History: Studies in Interdisciplinary History*. Princeton University Press, New Jersey.

Delworth T., and T.R. Knutson. 2000. Simulation of early 20th century global warming. *Science* 287:2246-2250.

Delworth T., and S. Manabe. 1989. The influence of soil wetness on near-surface atmospheric variability. *Journal of Climate* 2(12):1447-1462.

Delworth T., and M.E. Mann. 2000. Observed and simulated multidecadal variability in the Northern Hemisphere. *Climate Dynamics* 16(9):661-676.

deMenocal, P., J. Ortiz, T. Guilderson, J. Adkins, M. Sarnthein, L. Baker, and M. Yarusinsky. 2000a. Abrupt onset and termination of the African Humid Period: Rapid climate responses to gradual insolation forcing. *Quaternary Science Reviews* 19:347-361.

deMenocal, P., J. Ortiz, T. Guilderson, and M. Sarnthein. 2000b. Coherent high and low-latitude climate variability during the Holocene warm period. *Science* 288:2198-2202.

Denton, G.H., and C.H. Hendy. 1994. Documentation of an advance of New Zealand's Franz Joseph Glacier at the onset of Younger Dryas time. *Science* 264:1434-1437.

Denton, G.H., C.J. Heusser, T.V. Lowell, P.I. Moreno, B.G. Anderson, L.E. Heusser, C. Schluchter, and D.R. Marchant. 1999. Interhemispheric linkage of paleoclimate during the last deglaciation. *Geografiska Annaler* 81(2):107-153.

Diaz, H.F. 1983. Drought in the United States—Some aspects of major dry and wet periods in the contiguous United States, 1895-1981. *Journal of Climate and Applied Meteorology* 22:3-16.

Diaz, H.F., and V. Markgraf. 2000. *El Niño and the Southern Oscillation: Multiscale Variability and Global Impacts*. Cambridge University Press, UK. 496 pp.

Dickens, G.R. 1999. Carbon cycle: The blast in the past. *Nature* 401:752-755.

Dickens, G.R., J.R. O'Neil, D.K. Rea, and R.M. Owen. 1995. Dissociation of oceanic methane hydrate as a cause of the carbon-isotope excursion at the end of the Paleocene. *Paleoceanography* 10:965-971.

Dickson, R.R. 1999. All change in the Arctic. *Nature* 397:389-391.

Dickson, R.R. 2001. Centre for Environment, Fisheries, and Aquaculture Science, UK. Personal communication.

Dickson, R.R., J. Meincke, S.A. Malmberg, and A.J. Lee. 1988. The "Great Salinity Anomaly" in the northern North Atlantic 1968-1982. *Progress in Oceanography* 20:103-151.

Dickson, R.R., J. Lazier, J. Meincke, P. Rhines, and J. Swift. 1996. Long-term coordinated changes in the convective activity of the North Atlantic. *Progress in Oceanography* 38: 241-295.

Dickson, R.R., T.J. Osborn, J.W. Hurrell, J. Meincke, J. Blindheim, B. Adlandsvik, T. Vinje, G. Alekseev, and W. Maslowski. 2000. The Arctic Ocean response to the North Atlantic oscillation. *Journal of Climate* 13:2671-2696.

Dirmeyer, P.A. 1994. Vegetation stress as a feedback mechanism in midlatitude drought. *Journal of Climate* 7:1463-1483.

Dirmeyer, P.A., and J. Shukla. 1996. The effect on regional and global expansion of the world's deserts. *Quarterly Journal of the Royal Meteorological Society* 122:451-482.

Dixon, K.W., T.L. Delworth, M.J. Spelman, and R.J. Stouffer. 1999. The influence of transient surface fluxes on North Atlantic overturning in a coupled GCM climate change experiment. *Geophysical Research Letters* 26:2749-2752.

Doney, S.C., and W.J. Jenkins. 1994. Ventilation of the Deep Western Boundary Current and Abyssal Western North Atlantic: Estimates from tritium and He distributions. *Journal of Physical Oceanography* 24:638-659.

Donnelly, J.P., S.S. Bryant, J. Butler, J. Dowling, L. Fan, N. Hausmann, P. Newby, B. Shuman, J. Stern, K. Westover, and T. Webb. 2001a. Seven hundred year sedimentary record of intense hurricane landfalls in southern New England. *Geological Society of America Bulletin* 113:714-727.

Donnelly, J.P., S. Roll, M. Wengren, J. Butler, R. Lederer, and T. Webb. 2001b. Sedimentary evidence of intense hurricane strikes from New Jersey. *Geology* 29:615-618.

Döscher, R., C.W. Boening, and P. Herrmann. 1994. Response of circulation and heat transport in the North Atlantic to changes in thermohaline forcing in northern latitudes: A model study. *Journal of Physical Oceanography* 24:2303-2320.

Dunn, P.O., and D.W. Winkler. 1999. Climate change has affected the breeding date of tree swallows throughout North America. *Proceedings of the Royal Society* 266:2487-2490.

Duplessy, J.C., L. Labeyrie, M. Arnold, M. Paterne, J. Duprat, and T.C.E. Vanweering. 1992. Changes in surface salinity of the North Atlantic Ocean during the last deglaciation. *Nature* 358:485-488.

Durkee, P.A., K.J. Noone, and R.T. Bluth. 2000. The Monterey area ship track experiment. *Journal of Atmospheric Science* 57:2523-2541.

Durre, I.M. 2001. Research Scientist, National Oceanic and Atmospheric Administration, National Climate Data Center, Asheville, NC. Personal communication.

Dymnikov, V.P., and A.S. Gritsoun. 2001. Climate model attractor: Chaos, quasi-regularity and sensitivity to small perturbations of external forcing. *Nonlinear Processes in Geophysics* 8:201-209.

Dyurgerov, M.B., and M.F. Meier. 2000. Twentieth century climate change: Evidence from small glaciers. *Proceedings of the National Academy of Sciences of the United States of America* 97(4):1406-1411.

Easterling, D.R., G.A. Meehl, C. Parmesan, S.A. Changnon, T.R. Karl, and L.O. Mearns. 2000. Climate extremes: Observations, modeling, and impacts. *Science* 289:2068-2074.

Ebbesmeier, C.C., D.R. Cayan, D.R. Milan, F.H. Nichols, D.H. Peterson, and K.T. Redmond. 1991. 1976 step in the Pacific climate: Forty environmental changes between 1968-1975 and 1977-1984. In J.L. Betancourt and V.L. Tharp (eds.), *Proceedings of the Seventh Annual Pacific Climate Workshop*, April 1990. California Department of Water Resources, Interagency Ecological Studies Program, Technical Report 26.

Ekwurzel, B., P. Schlosser, R.A. Mortlock, R.G. Fairbanks, and J.H. Swift. 2001. River runoff, sea ice meltwater, and Pacific water distribution and mean residence times in the Arctic Ocean. *Journal of Geophysical Research* 106(C5):9075-9092.

Eltahir, E.A.B., and R.L. Bras. 1996. Precipitation recycling. *Review of Geophysics* 34: 367-378.

Ely, L. 1997. Response of extreme floods in the southwestern United States to climatic variations in the late Holocene. *Geomorphology* 19:175-201.

Ely, L., Y. Enzel, V.R. Baker, and D.R. Cayan. 1993. A 5000-year record of extreme floods and climate change in the southwestern United States. *Science* 262:410-412.

Entekhabi, D., I. Rodriguez-Iturbe, and R.L. Bras. 1992. Variability in large-scale water balance with land surface-atmosphere interaction. *Journal of Climate* 5:798-814.

Enzel, Y., L. Ely, P. House, and V. Baker. 1996. Magnitude and frequency of Holocene palaeofloods in the southwestern United States: A review and discussion of implications. In Branson, J., A. Brown, and K. Gregory (eds.), *Global Continental Changes: The Context of Palaeohydrology*, Vol. 115. Geological Society Special Publication, London.

Enzel, Y., L. Ely, S. Mishra, R. Ramesh, R. Amit, B. Lazar, S.N. Rajaguru, V.R. Baker, and A. Sandler. 1999. High-resolution Holocene environmental changes in the Thar Desert, northwestern India. *Science* 284:125-128.

Epstein, P.R. 1998. Global warming and vector-borne disease. *Lancet* 351:1737.

Fahnestock, M., R. Bindschadler, R. Kwok, and K. Jezek. 1993. Greenland ice sheet surface properties and ice dynamics form ERS-1 SAR imagery. *Science* 262:1530-1534.

Fanning, A.F., and A.J. Weaver. 1997. Temporal-geographical influences on the North Atlantic conveyor: Implications for the Younger Dryas. *Paleoceanography* 12:307-320.

Fawcett, P.J., A.M. Águstsdóttir, R.B. Alley, and C.A. Shuman. 1997. The Younger Dryas termination and North Atlantic deep water formation: Insights from climate model simulations and Greenland ice cores. *Paleoceanography* 12(1):23-38.

Federov, A.V., and S.G. Philander. 2000. Is El Niño changing? *Science* 288:1997-2001.

Fennessy, M.J. and J. Shukla. 1999. Impact of initial soil wetness on seasonal atmospheric prediction. *Journal of Climate* 12:3167-3180.

Findell, K.L., and E.A.B. Eltahir. 1999. Analysis of the pathways relating soil moisture and subsequent rainfall in Illinois. *Journal of Geophysical Research* 104:31,165-31,174.

Fisher, D.A., R.M. Koerner, and N. Reeh. 1995. Holocene climate records frm Agassiz Ice Cap, Ellesmere Island, NNT, Canada. *Holocene* 5:19-24.

Forman, S.L., R. Oglesby, V. Markgraf, and T. Stafford. 1995. Paleoclimatic significance of late Quaternary eolian deposition on the Piedmont and High-Plains, Central United-States. *Global and Planetary Change* 11:35-55.

Forman, S.L., R. Oglesby, and R.S. Webb. 2001. Temporal and spatial patterns of Holocene dune activity on the Great Plains of North America: megadroughts and climate links. *Global and Planetary Change* 29(1-2):1-29.

Francis, R.C., and S.R. Hare. 1994. Decadal-scale regime shifts in the large marine ecosystems of the northeast Pacific: A case for historical science. *Fisheries Oceanography* 3:279-291.

Fritz, S.C., S.E. Metcalfe, and W. Dean. 2000. Holocene climate patterns in the Americas inferred from paleolimnological records. Pp. 241-259 in V. Markgraf (ed.), *Interhemispheric Climate Linkages*. Academic Press, New York.

Fronval, T., E. Jansen, H. Haflidason, and H.P. Sejrup. 1998. Variability in surface and deep water conditions in the Nordic Seas during the last interglacial period. *Quaternary Science Reviews* 17(9-10):963-985.

Funder S., and A. Weidick. 1991. Holocene Boreal Mollusks In Greenland - Palaeoceanographic Implications. *Palaeogeography Palaeoclimate* 85 (1-2): 123-135

Fyfe, J.C., G.J. Boer, and G.M. Flato. 1999. The Arctic and Antarctic oscillations and their projected changes under global warming. *Geophysical Research Letters* 26:1601-1604.

Gammelsrød T., S. Østerhus, and Ø. Godøy. 1992. Decadal variations of ocean climate in the Norwegian Sea observed at Ocean Station "Mike" (65°N 2°E). *ICES Marine Symposium* 195:68-75.

Ganachaud, A., and C. Wunsch. 2000. Improved estimates of global ocean circulation, heat transport and mixing from hydrographic data. *Nature* 408:453-457.

Ganopolski, A., C. Kubatzki, M. Claussen, V. Brovkin, and V. Petoukhov. 1998. The influence of vegetation-atmosphere-ocean interaction on climate during the mid-Holocene. *Science* 280:1916-1919.

Ganopolski, A., and S. Rahmstorf. 2001. Rapid changes of glacial climate simulated in a coupled climate model. *Nature* 409:153-158.

Gardner, B.R., B.L. Blad, and D.G. Watts. 1981. Plant and air temperatures in differentially irrigated corn. *Agricultural Meteorology* 25:207-217.

Gasse, F. 2000. Hydrological changes in the African tropics since the Last Glacial Maximum. *Quaternary Science Reviews* 19:189-211.

Gasse, F., and E. van Campo. 1994. Abrupt post-glacial climate events in West Asia and North Africa monsoon domains. *Earth and Planetary Science Letters* 126:435-456.

Gent, P.R. 2001. Will the North Atlantic Ocean thermohaline circualtion weaken during the 21st century? *Geophysical Research Letters* 28:1023-1026.

Georgakakos, K.P., D.-H. Bae, and D.R. Cayan. 1995. Hydroclimatology of continental watersheds. I. Temporal analysis. *Water Resources Research* 31:655-675.

Gill, R.B. 2000. *The Great Maya Droughts: Water, Life and Death*. University of New Mexico Press, Albuquerque.

Glantz, M. 1999. *Creeping Environmental Problems and Sustainable Development in the Aral Sea Basin*. Cambridge University Press, UK.

Gleick, P.H. 2000. *The World's Water 2000-2001*. Island Press, Washington, D.C. 315 pp.

Gnanadesikan, A. 1999. A simple predictive model for the structure of the oceanic pycnocline. *Science* 283:2077-2079.

Goñi, M.F.S., F. Eynaud, J.L. Turon, and N.J. Shackleton. 1999. High resolution palynological record off the Iberian margin: Direct land-sea correlation for the Last Interglacial complex. *Earth and Planetary Science Letters* 171(1):123-137.

Gordon, A.L. 1986. Interocean exchange of thermocline water. *Journal of Geophysical Research* 91:5037-5046.

Gordon, A.L., and J. C. Comiso. 1988. Polynyas in the Southern Ocean. *Scientific American* 256(6):90-97.

Gorham, E. 1991. Northern peatlands: Role in the carbon cycle and probable responses to climatic warming. *Ecological Applications* 1:182-195.

Gornitz V., C. Rosenzweig, and D. Hillel. 1997. Effects of anthropogenic intervention in the land hydrologic cycle on global sea level rise. *Global and Planetary Change* 14:147-161.

Gosse, J.C., J. Klein, E.B. Evenson, B. Lawn, and R. Middleton. 1995. Beryllium-10 dating of the duration and retreat of the last Pinedale glacial sequence. *Science* 268:1329-1333.

Graham, N.E. 1994. Decadal-scale climate variability in the tropical and North Pacific during the 1970s and 1980s: Observations and model results. *Climate Dynamics* 10:135-162.

Gregory, K.J., L. Starkel, and V.R. Baker. 1995. *Global Continental Palaeohydrology*. John Wiley & Sons, Ltd., New York. 334 pp.

Grootes, P.M., and M. Stuiver. 1997. Oxygen 18/16 variability in Greenland snow and ice with 10^{-3}- to 10^5-year time resolution. *Journal of Geophysical Research* 102:26,455-26,470.

Grousset, F.E., L. Labeyrie, J.A. Sinko, M. Cremer, G. Bond, J. Duprat, E. Cortijo, and S. Huon. 1993. Patterns of ice-rafted detritus in the glacial North Atlantic (40-55°N). *Paleoceanography* 8:175-192.

Guay, C.K.H., K.K. Falkner, R.D. Muench, M. Mensch, M. Frank, and R. Bayer. 2001. Wind-driven transport pathways for Eurasian Arctic river discharge. *Journal of Geophysical Research* 106(2):11,469-11,480.

Guilderson T.P., and D.P. Schrag. 1998. Abrupt shift in subsurface temperatures in the tropical Pacific associated with changes in El Niño. *Science* 281(5374):240-243.

Gwiazda, R.H., S.R. Hemming, and W.S. Broecker. 1996a. Provenance of icebergs during Heinrich event 3 and the contrast to their sources during other Heinrich episodes. *Paleoceanography* 11:371-78.

Gwiazda, R.H., S.R. Hemming, W.S. Broecker, T. Onsttot, and C. Mueller. 1996b. Evidence from ^{40}Ar/^{39}Ar ages for a Churchill province source of ice-rafted amphiboles in Heinrich layer 2. *Journal of Glaciology* 42:440-46.

Häkkinen, S. 1993. An Arctic source for the Great Salinity Anomaly: A simulation of the Arctic ice-ocean system for 1955-1975. *Journal of Geophysical Research* 98:16,397-16,410.

Häkkinen, S. 1999. Variability of the simulated meridional heat transport in the North Atlantic for the period 1951-1993. *Journal of Geophysical Research* 104:10,991-11,007.

Häkkinen, S. Submitted. Variability in sea-surface height: A qualitative measure for the meridional overturning in the North Atlantic. *Journal of Geophysical Research*.

Hall, A., and R.J. Stouffer. 2001. An abrupt climate event in a coupled ocean-atmosphere simulation without external forcing. *Nature* 409:171-174.

Hansen, A.R., and A. Sutera. 1995. The probability density distribution of the planetary-scale atmospheric wave amplitude revisited. *Journal of Atmospheric Sciences* 52:2463-2472.

Hansen, A.J., R.R. Neilson, V.H. Dale, C.H. Flather, L.R. Iverson, D.J. Currie, S. Shafer, R. Cook, and P.J. Bartlein. 2001a. Global change in forests: Responses of species, communities, and biomes. *Bioscience* 51(9):765-779.

Hansen, B., W.R. Turrell, and S. Osterhus. 2001b. Decreasing overflow from the Nordic seas into the Atlantic Ocean through the Faroe Bank channel since 1950. *Nature* 411:927-930.

Harland, W.B. 1964. Critical evidence for a great infra-Cambrian glaciation. *Geologische Rundschau* 54:45-61.

Hawkins, B.A., and M. Holyoak. 1998. Transcontinental crashes of insect populations? *American Naturalist* 152:480-484.

Hegerl, G.C., P.A. Stott, M.R. Allen, J.F.B. Mitchell, S.F.B. Tett, and U. Cubasch. 2000. Optimal detection and attribution of climate change: Sensitivity of results to climate model differences. *Climate Dynamics* 16:737-754.

Heinrich H. 1988. Origin and consequences of cyclic ice rafting in the northeast Atlantic Ocean during the past 130,000 years. *Quaternary Research* 29:143-152.

Held, I.M., and B.J. Soden. 2000. Water vapor feedback and global warming. *Annual Review of Energy and the Environment* 25:441-475.

Henderson-Sellers, A., K. McGuffie, and C. Gross. 1995. Sensitivity of global climate model simulations to increased stomatal resistance and CO_2 increases. *Journal of Climate* 8:1735-1756.

Hepting, G.H. 1963. Climate and forest disease. *Annual Review of Phytopathology* 1:31-50.

Herman, S.W. 2000. Fixed assets and consumer durable goods. *Survey of Current Business*. U.S. Department of Commerce, Bureau of Economic Analysis, Washington, D.C. 17 pp.

Heusser, C.J. 1990. Late-glacial and Holocene vegetation and climate of sub-Antarctic South-America. *Review of Paleobotany and Palynology* 65(1-4):9-15.

Hillaire-Marcel, C., A. deVernal, G. Bilodeau, and A.J. Weaver. 2001. Absence of deep-water formation in the Labrador Sea during the last interglacial period. *Nature* 410:1073-1077.

Hodell, D.A., J.H. Curtis, and M. Brenner. 1995. Possible role of climate in the collapse of Classic Maya civilization. *Nature* 375:391-394.

Hodell, D.A., M. Brenner, J.H. Curtis, and T. Guilderson. 2001. Solar forcing of drought frequency in the Maya lowlands. *Science* 292:1367-1370.

Hoerling, M.P., J.W. Hurrell, and T. Xu. 2001. Tropical origins for recent North Atlantic climate change. *Science* 292:90-92.

Hoffman, P.F., A.J. Kaufman, G.P. Halverson, and D.P. Schrag. 1998. A Neoproterozoic snowball Earth. *Science* 281:1342-1346.

Hooghiemstra, H., A.M. Cleef, G.W. Noldus, and M. Kappell. 1992. Upper Quaternary vegetation dynamics and paleoclimatology of the La Chonta bog area (Cordillera de Talamanca, Costa Rica). *Journal of Quaternary Science* 7:205-225.

Hostetler, S.W., P.U. Clark, P.J. Bartlein, A.C. Mix, and N.J. Pisias. 1999. Atmospheric transmission of North Atlantic Heinrich events. *Journal of Geophysical Research Atmospheres* 104(D4):3947-3952.

Hou A.Y., and R.S. Lindzen. 1992. The influence of concentrated heating on the Hadley circulation. *Journal of Atmospheric Sciences* 49(14):1233-1241.

Huang, J., and H.M. Van den Dool. 1993. Monthly precipitation-temperature relations and temperature predictions over the United States. *Journal of Climate* 6:1111-1132.

Huang, J., H.M. Van den Dool, and K.P. Georgakakos. 1996. Analysis of model-calculated soil moisture over the United States (1931-1993) and applications to long-range temperature forecasts. *Journal of Climate* 9:1350-1362.

Huang, S., H. Pollack, and P.Y. Shen. 2000. Temperature trends over the past five centuries reconstructed from borehole temperatures. *Nature* 403:756-758.

Hughen, K.A., J.T. Overpeck, L.C. Peterson, and S. Trumbore. 1996. Rapid climate changes in the tropical Atlantic region during the last deglaciation. *Nature* 380:51-54.

Hughen, K.A., D.P. Schrag, S.B. Jacobsen, and W. Hantoro. 1999. El Niño during the last interglacial period recorded by a fossil coral from Indonesia. *Geophysical Research Letters* 26(20):3129-3132.

Hughen, K.A., J.T. Overpeck, and R.F. Anderson. 2000a. Recent warming in a 500-year palaeotemperature record from varved sediments, Upper Soper Lake, Baffin Island, Canada. *The Holocene* 10:9-19.

Hughen, K.A., J.R. Southon, S.J. Lehman, and J.T. Overpeck. 2000b. Synchronous radiocarbon and climate shifts during deglaciation. *Science* 290:1951-1954.

Hughen, K.A., J.T. Overpeck, S.J. Lehman, M. Kashgarian, J. Southon, L.C. Peterson, R. Alley and D.M. Sigman. 1998. Deglacial changes in ocean circulation from an extended radiocarbon calibration. *Nature* 391:65-68.

Hulbe, C.L., and A.J. Payne. 2001. The contribution of numerical modelling to our understanding of the West Antarctic ice sheet. Pp. 201-219 in *The West Antarctic Ice Sheet: Behavior and Environment*, Antarctic Research Series 77. American Geophysical Union, Washington, D.C .

Hurrell, J.W. 1995. Decadal trends in the North Atlantic Oscillation: Regional temperatures and precipitation. *Science* 269:676-679.

Hyde, W.T., T.J. Crowley, S.K. Baum, and W.R. Peltier. 2000. Neoproterozoic 'snowball Earth' simulations with a coupled climate/ice-sheet model. *Nature* 405(6785):425-429.

Inouye, D.W., and A.D. McGuire. 1991. Effects of snowpack on timing and abundance of flowering in *Delphinium nelsonii* (*Ranunculaceae*): Implications for climatic change. *American Journal of Botany* 78:997-1001.

Inouye, D.W. 2000. The ecological and evolutionary significance of frost in the context of climate change. *Ecology Letters* 3(5):457-463.

Inouye, D.W., B. Barr, K.B. Armitage, and B.D. Inouye. 2000. Climate change is affecting altitudinal migrants and hibernating species. *Proceedings of the National Academy of Sciences of the United States of America* 97:1630-1633.

Intergovernmental Panel on Climate Change. 2001a. *Climate Change 2001: Impacts, Adaptations, and Vulnerability.* Report of Working Group II. Available online at: http://www.usgcrp.gov/ipcc/default.html [2001, October 25].

Intergovernmental Panel on Climate Change. 2001b. *Climate Change 2001: The Science of Climate Change,* J.T. Houghton (ed.). Cambridge University Press, UK.

International Red River Basin Task Force. 2000. The Next Flood: Getting Prepared, Final Report of the International Red River Basin Task Force to the International Joint Commission, April 2000, International Joint Commission. Available online at: http://www.ijc.org/boards/rrbtf.html

Iversen, J. 1954. The late-glacial flora of Denmark and its relation to climate and soil. *Damn. Geol. Unders. Ser. II* 80:87-119.

Jacka, T.H., and W.F. Budd. 1998. Detection of temperature and sea-ice-extent changes in the Antarctic and Southern Ocean, 1949-96. *Annals of Glaciology* 27:553-559.

Jansen, K. 1938. Some west Baltic pollen diagrams. *Quartar* 1:124-139.

Jennings, A.E., and N.J. Weiner. 1996. Environmental change in eastern Greenland during the last 1300 years: Evidence from foraminifera and lithofacies in Nansen Fjord, 68°N. *Holocene* 6:179-191.

Jennings, A.E., K.L. Knudsen, M. Hald, C.V. Hansen, and J.T. Andrews. In press. A mid-Holocene shift in Arctic sea ice variability on the East Greenland Shelf. *The Holocene.*

Johannessen, O.M., E.V. Shalina, and M.W. Miles. 1999. Satellite evidence for an Arctic sea ice cover in transformation. *Science* 286:1937-1939.

Johnsen, S.J., H.B. Clausen, W. Dansgaard, N.S. Gundestrup, C.U. Hammer, U. Andersen, K.K. Andersen, C.S. Hvidberg, D. Dahl-Jensen, J.P. Steffensen, H. Shoji, A.E. Sveinbjornsdottir, J. White, J. Jouzel, and D. Fisher. 1997. The δ^{O-18} record along the Greenland Ice Core Project deep ice core and the problem of possible Eemian climatic instability. *Journal of Geophysical Research* 102(26):397-410.

Johnsen, S.J., D. Dahl-Jensen, W. Dansgaard, and N. Gundestrup. 1995. Greenland paleo-temperatures derived from GRIP bore hole temperature and ice core isotope profiles. *Tellus* 47B:624-629.

Johnson, G.C., and A.H. Orsi. 1997. Southwest Pacific Ocean water-mass changes between 1968/69 and 1990/91. *Journal of Climate* 10:306-316.

Jones, P.D., and R.S. Bradley. 1992. Climatic variations over the last 500 years. Pp. 649-665 in R.S. Bradley and P.D. Jones (eds.), *Climate Since A.D. 150.* Routledge, London.

Jones, C.G., R.S. Ostfeld, M.P. Richard, E.M. Schauger, and J.O. Wolff. 1998. Chain reactions linking acorns to gypsy moth outbreaks and Lyme disease risk. *Science* 279:1023-1026.

Karl, T. 1983. Some spatial characteristics of drought in the United States. *Journal of Climate and Applied Meteorology* 22:1356-1366.

Karl, T. 1986. Relationships Between Some Moisture Parameters and Subsequent Seasonal and Monthly Mean Temperature in the United States. Programme on Long-Range Forecasting Res. Rep. Ser., No. 6, WMO/TDD-No. 87, 2, 661-670. [Available from WMO; Office for Information and Public Affairs; 41, Avenue Giuseppe-Motta; 1211 Geneva 2 / Switzerland; PubSales@gateway.wmo.ch.]

Karl, T., and R.G. Quayle. 1981. The 1980 summer heat wave and droughts in historical perspective. *Monthly Weather Review* 109:2055-2073.

Kates, R.W., B.L. Turner II, and W.C. Clark. 1990. The great transformation. Pp. 1-17 in B.L. Turner II, W.C. Clark, R.W. Kates, J.F. Richards, J.T. Mathews, and W.B. Meyer (eds.), *The Earth as Transformed by Human Action*. Cambridge University Press, New York.

Kawase, M. 1987. Establishment of deep ocean circulation driven by deep water production. *Journal of Physical Oceanography* 17:2294-2317.

Keigwin, L.D. 1996. The Little Ice Age and Medieval Warm Period in the Sargasso Sea. *Science* 274:1504-1508.

Keigwin, L.D., and G.A. Jones. 1990. Deglacial climatic oscillations in the Gulf of California. *Paleoceanography* 5:1009-1023.

Keller, K. 2000. *Trace Metal-Carbon Interactions in Marine Phytoplankton: Implications on the Cellular, Regional, and Global Scale*. Ph.D. thesis, Princeton University. Also available online at: http:\\geoweb.princeton.edu\bigscience\GLODAP\details.html\#Redfield [April 25, 2001].

Keller, K., K. Tan, F.M.M. Morel, and D.F. Bradford. 2000. Preserving the ocean circulation: Implications for climate policy. *Climatic Change* 47(1-2):17-43.

Keller, K., B.M. Bolker, and D.F. Bradford. 2001. *Uncertain Climate Thresholds in Economic Optimal Growth Models*. Yale/NBER Workshop on the Societal Impact of Abrupt Climate Change, Snowmass, Colorado, July 24-25.

Kelley, P.M., C.S. Goodess, and B.S.G. Cherry. 1987. The interpretation of the Icelandic sea ice record. *Journal of Geophysical Research* 92:10,835-10,843.

Kelley, P.M., P.D. Jones, C.B. Sear, B.S.G. Cherry, and R.K. Tavakol. 1982. Variations in surface air temperature: Part 2, Arctic regions. *Monthly Weather Review* 110:71-83.

Kennett, J.P., and L.D. Stott. 1991. Abrupt deep-sea warming, palaeoceanographic changes and benthic extinctions at the end of the Paleocene. *Nature* 353:225-229.

Kennett, J.P., and B.L. Ingram. 1995. A 20,000-year record of ocean circulation and climate change from the Santa Barbara basin. *Nature* 377:510-513.

Kennett J.P., K.G. Cannariato, I.L. Hendy, and R.J. Behl. 2000. Carbon isotopic evidence for methane hydrate instability during Quaternary interstadials. *Science* 288(5463):128-133.

Kirschvink, J.L. 1992. Late Proterozoic low-latitude global glaciation: The Snowball Earth. Pp. 51-52 in J.W. Schopf and C. Klein (eds.), *The Proterozoic Biosphere*. Cambridge University Press, New York.

Kitzberger, T., T.W. Swetnam, and T.T. Veblen. 2001. Inter-hemispheric synchrony of forest fires and the El Niño-Southern Oscillation. *Global Ecology and Biogeography* 10(3):315-326.

Kling, G.W. 2001. *The Response of Freshwater Ecosystems to Abrupt Climate Change*. Summary notes from the Abrupt Climate Change Meeting, National Academy of Sciences, March 21-23, 2001, Washington, D.C.

Klinger, B.A., and J. Marotzke. 1999. Behavior of double-hemisphere thermohaline flows in a single basin. *Journal of Physical Oceanography* 29:382-400.

Kneller, M.D., and D.M. Peteet. 1999. Late-glacial to early Holocene climate changes from a central Appalachian pollen and macrofossil record. *Quaternary Research* 51:133-147.

Knox, J.C. 1999. Long-term episodic changes in magnitudes and frequencies of floods in the Upper Mississippi River Valley. In A.G. Brown, and T.A. Quine (eds.), *Fluvial Processes and Environmental Change*. John Wiley and Sons, Ltd., London.

Knox, J.C. 2000. Sensitivity of modern and Holocene floods to climate change. *Quaternary Science Reviews* 19:439-457.

Knutti, R., and T.F. Stocker. 2000. Influence of the thermohaline circulation on projected sea level rise. *Journal of Climate* 13:1997-2001.

Knutti, R., and T.F. Stocker. 2001. Limited predictability of the thermohaline circulation close to a threshold. *Journal of Climate* 15:179-186.

Koenig, W.D., and J.M.H. Knops. 2000. Patterns of annual seed production by Northern Hemisphere trees: a global perspective. *The American Naturalist* 155:59-69.

Koster, R.D., and M.J. Suarez. 1996. The influence of land surface moisture retention on precipitation statistics. *Journal of Climate* 9:2551-2567.

Kotilainen, A.T., and N.J. Shackleton. 1995. Rapid climate variabilitly in the North Pacific Ocean during the past 95,000 years. *Nature* 377:323-326.

Koutavas, A., and J. Lynch-Steiglitz. 1999. *A Younger Dryas Temperature Reversal in the Eastern Equatorial Pacific.* Presented at the American Geophysical Union Fall Meeting. American Geophysical Union, Washington, D.C.

Krabill, W., W. Abdalati, E. Frederick, S. Manizade, C. Martin, J. Sonntag, R. Swift, R. Thomas, W. Wright, and J. Yungel. 2000. Greenland ice sheet: High-elevation balance and peripheral thinning. *Science* 289:428-430.

Kudo, G., U. Nordenhall, and U. Molau. 1999. Effects of snowmelt timing on leaf traits, leaf production, and shoot growth of alpine plants: Comparisons along a snowmelt gradient in northern Sweden. *Ecoscience* 6:439-450.

Kutzbach, J.E., and Z. Liu. 1997. Response of the African monsoon to orbital forcing and ocean feedbacks in the middle Holocene. *Science* 278:440-443.

Kutzbach, J., G. Bonan, J. Foley, and S. Harrison. 1996. Vegetation and soil feedbacks on the response of the African monsoon to orbital forcing in the early to middle Holocene. *Nature* 384:623-626.

Kwok, R. G.F. Cunningham, and S. Yueh. 1999. Area balance of the Arctic Ocean perennial ice zone: October 1996 to April 1997. *Journal of Geophysical Research* 104(C11):25,747-25,759.

Laird, K.R., S.C. Fritz, K.A. Maasch, and B.F. Cumming. 1996. Greater drought intensity and frequency before AD 1200 in the Northern Great Plains, USA. *Nature* 384:552-555.

Lamb, H.F., F. Gasse, A. Benkaddour, N. El Hamouti, S. van der Kaars, W.T. Perkins, N.J. Pearce, and C.N. Roberts. 1995. Relation between century-scale Holocene arid intervals in tropical and temperate zones. *Nature* 373:134-137.

Lang, C., M. Leuenberger, J. Schwander, and S. Johnsen. 1999. 16 degrees C rapid temperature variation in Central Greenland 70,000 years ago. *Science* 286(5441):934-937.

Lare, A.R. and S.E. Nicholson. 1994. Contrasting conditions of land surface atmosphere feedback mechanism in the West African Sahel. *Journal of Climate* 7:653-668.

Latif, M., and T.P. Barnett. 1996. Decadal climate variability over the North Pacific and North America: Dynamics and predictability. *Journal of Climate* 9:2407-2423.

Latif, M., E. Roeckner, U. Mikolajewicz, and R. Voss. 2000. Tropical stabilization of the thermohaline circulation in a greenhouse warming simulation. *Journal of Climate* 13:1809-1813.

Lawton, J. H., and R.M. May (eds.). 1995. *Extinction Rates.* Oxford University Press, UK.

Lazier, J.R.N. 1995. The salinity decrease in the Labrador Sea over the past 30 years. Pp. 295-304 in D.G. Martinson, K. Bryan, M. Ghil, M.M. Hall, T.R. Karl, E.S. Sarachik, S. Sorooshian, and L.D. Talley (eds.), *Natural Climate Variability on Decade-to-Century Time Scales.* National Academy Press, Washington D.C.

Lee, A. 1949. *Reun. Cons. Int. Explor. Mer.* 125:43-52.

Lee, K.E., N.C. Slowey, T.D. Herbert. 2001. Glacial sea surface temperatures in the subtropical North Pacific: A comparison of U-k(37)', $\delta^{18}O$, and foraminiferal assemblage temperature estimates. *Paleoceanography* 16(3):268-279.

Lee, T., and J. Marotzke. 1997. Inferring meridional mass and heat transports of the Indian Ocean by fitting a general circulation model to climatological data. *Journal of Geophysical Research* 102(10):585-602.

Lehman, S.J., and L.D. Keigwin. 1992. Sudden changes in the North Atlantic circulation during the last deglaciation. *Nature* 356:757-762.

Leopold, C. 2001. *Biological Response to Abrupt Climate Change.* Summary notes from the Abrupt Climate Change meeting, National Academy of Sciences, March 21-23, 2001, Washington, D.C.

Levitus, S., J.I. Antonov, J. Wang, T.L. Delworth, K.W. Dixon, and A.J. Broccoli. 2001. Anthropogenic warming of Earth's climate system. *Science* 292:267-270.

Levitus, S., J.I. Antonov, T.P. Boyer, and C. Stevens. 1999. Warming of the world ocean. *Science* 287:2226-2229.

Levitus S. 2001. National Oceanic and Atmospheric Administration. Personal communication. Social Impacts Workshop.

Leyden, B.W. 1995. Evidence of the Younger Dryas in Central America. *Quaternary Science Reviews* 14(9):833-840.

Leyden, B.W., M. Brenner, D.A. Hodell, and J.H. Curtis. 1994. Orbital and internal forcing of climate on the Yucatan Peninsula for the past 36kyr. *Palaeogeography, Palaeoclimatology, Palaeoecology* 109:193-210.

Lindzen, R.S., and A.Y. Hou. 1988. Hadley circulations for zonally averaged heating centered off the equator. *Journal of the Atmospheric Sciences* 45:2416-2427.

Lilly, J.M., P.B. Rhines, M. Visbeck, R. Davis, J.R.N. Lazier, F. Schott, and D. Farmer. 1999. Observing deep convection in the Labrador sea during winter 1994/95. *Journal of Physical Oceanography* 29(8):2065-2098.

Linthicum K.J., A. Anyamba, C.J. Tucker, P.W. Kelley, M.F. Myers, and C.J. Peters. 1999. Climate and satellite indicators to forecast Rift Valley fever epidemics in Kenya. *Science* 285:397-400.

Liu, K.B., and M.L. Fearn. 2000. Reconstruction of prehistoric landfall frequencies of catastrophic hurricanes in northwestern Florida from lake sediment records. *Quaternary Research* 54:238-245.

Lorenz, E.N. 1963. Deterministic non-periodic flow. *Journal of the Atmospheric Sciences* 20:130-141.

Lorenz, E.N. 1990. Can chaos and intransitivity lead to interannual variability? *Tellus* 42(A):378-389.

Lowe, J.J., and C. Watson.1993. Late glacial and early Holocene pollen stratigraphy of the northern Apennines, Italy. *Quaternary Science Reviews* 12:727-738.

Lowe, J.J., T. Alm, B. Ammann, S. Anderson, T. Anderson, J.T. Andrews, W.E.N. Austin, J.L. Debeaulieu, B. Becker, K.E. Behre, B.E. Berglund, H. Bersten, H.J. Beug, H.H. Birks, S. Bjorck, S. Bohncke, A. Brande, L. Brunnberg, G.R. Coope, P. Coxon, L. Cwynar, L. Denys, J.J. Donner, R.A. Dugmore, J.C. Duplessy, H. Faure, M.J. Gaillard, I. Hajdas, D.D. Harkness, A. Hjartarson, W. Hoek, B. Huntley, O. Ingolfsson, G. Jacobsen, E. Jansen, E. Kolstrup, P. Kiden, N. Koc, B. Kromer, G. Jalut, J. Landvik, G. Lemdahl, A.J. Levesque, A.F. Lotter, J. Lundqvist, J. Mangerud, F.E. Mayle, U. Miller, R. Mott, H. Norddahl, M. O'Connell, A. Paus, D.M. Peteet, A. Pons, M. Reille, P. Richard, M. Saarnisto, J.I. Svendsen, K. Thors, H. Usinger, J.M. Vanmourik, Y. Vasari, C. Verbruggen, I. Walker, M.J.C. Walker, W.A. Watts, and B. Wohlfarth. 1995. Palaeoclimate of the Northern Atlantic seaboards during the lastglacial/interglacial transition. *Quaternary International* 28:51-62.

Lowell, T.V. 2000. As climate changes, so do glaciers. *Proceedings of the National Academy of Sciences of the United States of America* 97(4):1351-1354.

Lu P., J.P. McCreary, and B.A. Klinger. 1998. Meridional circulation cells and the source waters of the Pacific Equatorial Undercurrent. *Journal of Physical Oceanography* 28(1):62-84.

Lu P., and J.P. McCreary. 1995. Influence of the ITCZ on the flow of thermocline water from the subtropical to the equatorial Pacific Ocean. *Journal of Physical Oceanography* 25(12):3076-3088.

Lyle M. 1988. Climatically forced organic carbon burial in equatorial Atlantic and Pacific Oceans. *Nature* 335:529-32.

MacAyeal, D.R. 1992. Irregular oscillations of the West Antarctic ice sheet. *Nature* 359:29-32.

MacAyeal, D.R. 1993a. A low-order model of growth/purge oscillations of the Laurentide ice sheet. *Paleoceanography* 8:767-773.

MacAyeal, D.R. 1993b. Binge/purge oscillations of the Laurentide ice sheet as a cause of the North Atlantic's Heinrich events. *Paleoceanography* 8:775-784.

Macdonald, R.W., E.C. Carmack, F.A. McLaughlin, K.K. Falkner, and J.H. Swift. 1999. Connections among ice, runoff and atmospheric forcing in the Beaufort Gyre. *Geophysical Research Letters* 26(15):2223-2226.

Mahowald, N., K. Kohfeld, M. Hansson, Y. Balkanski, S.P. Harrison, I.C. Prentice, M. Schulz, and H. Rodhe. 1999. Dust sources and deposition during the last glacial maximum and current climate: A comparison of model results with paleodata from ice cores and marine sediments. *Journal of Geophysical Research* 104:15,895-1,5916.

Manabe, S., and A.J. Broccoli. 1985. The influence of continental ice sheets on the climate of an ice age. *Journal of Geophysical Research* 90(C2):2167-2190.

Manabe, S., and K. Bryan. 1985. CO_2-induced change in a coupled ocean-atmosphere model and its paleoclimatic implications. *Journal of Geophysical Research* 90:1689-1707.

Manabe, S., and R.J. Stouffer. 1988. Two stable equilibria of a coupled ocean atmosphere model. *Journal of Climate* 1:841-866.

Manabe, S., and R.J. Stouffer. 1993. Century-scale effects of increased atmospheric CO_2 on the ocean-atmosphere system. *Nature* 364:215-218.

Manabe, S., and R.J. Stouffer. 1994. Multiple-century response of a coupled ocean-atmosphere model to an increase of atmospheric carbon dioxide. *Journal of Climate* 7:5-23.

Manabe, S., and R.J. Stouffer. 1995. Simulation of abrupt climate change induced by freshwater input to the North Atlantic Ocean. *Nature* 378:165-167.

Manabe, S., and R.J. Stouffer. 1997. Coupled ocean-atmosphere model response to freshwater input: comparison to the Younger Dryas event. *Paleoceanography* 12:321-336.

Manabe, S., and R.J. Stouffer. 1999. Are two modes of the thermohaline circulation stable? *Tellus* 51(A):400-411.

Mangerud, J. 1991. The last interglacial/glacial cycle in Northern Europe. In L.C.K. Shane and E.J. Cushing (eds.), *Quaternary Landscapes*. University of Minnesota Press, Minneapolis.

Manion, P.D. 1991. *Tree Disease Concepts*. Prentice Hall Career and Technology, NJ. 402 pp.

Mann, M., R. Bradley, and M. Hughes. 1998. Global-scale temperature patterns and climate forcing over the past six centuries. *Nature* 392:779-787.

Mann, M., R. Bradley, and M. Hughes. 1999. Northern Hemisphere temperatures during the past millennium: Inferences, uncertainties, and limitations. *Geophysical Research Letters* 26:759-762.

Mann, M.E., E. Gille, R.S. Bradley, M.K. Hughes, J. Overpeck, F.T. Keimig, and W. Gross. 2000. Global temperature patterns in past centuries: An interactive presentation. *Earth Interactions* 4(4):1. Also available online at http://earthinteractions.org.

Mantua, N.J., S.R. Hare, Y. Zhang, J.M. Wallace, and R.C. Francis. 1997. A Pacific interdecadal climate oscillation with impacts on salmon production. *Bulletin of the American Meteorological Society* 78:1069-1079.

Marchitto, Jr., T.M., W.B. Curry, and D.W. Oppo. 1998. Millennial-scale changes in North Atlantic circulation since the last deglaciation. *Nature* 393:557-561.

Markgraf, V. 1991. Younger Dryas in Southern South-America. *Boreas* 20(63):63-69.

Marotzke, J. 1990. Instabilities and multiple equilibria of the thermohaline circulation. *Berichte aus dem Institut für Meereskunde, Kiel.* Ph.D. thesis. 126 pp.

Marotzke, J. 1996. Analysis of thermohaline feedbacks. In D.L.T. Anderson, and J. Willebrand (eds.), *Decadal Climate Variability: Dynamics and Predictability*, Series I, Vol. 44. Springer-Verlag, Berlin.

Marotzke, J. 1997. Boundary mixing and the dynamics of three-dimensional thermohaline circulations. *Journal of Physical Oceanography* 27:1713-1728.

Marotzke, J. 2000. Abrupt climate change and thermohaline circulation: Mechanisms and predictability. *Proceedings of the National Academy of Sciences of the United States of Ameria* 97:1347-1350.

Marotzke, J., and J. Willebrand. 1991. Multiple equilibria of the global thermohaline circulation. *Journal of Physical Oceanography* 21:1372-1385.

Marotzke, J., and J.R. Stott. 1999. Convective mixing and the thermohaline circulation. *Journal of Physical Oceanography* 29:2962-2970.

Marotzke, J., and B.A. Klinger. 2000. The dynamics of equatorially asymmetric thermohaline circulations. *Journal of Physical Oceanography* 30:955-970.

Marotzke, J., P. Welander, and J. Willebrand. 1988. Instability and multiple equilibria in a meridional-plane model of the thermohaline circulation. *Tellus* 40A:162-172.

Marshall, J., and F. Schott. 1999. Open-ocean convection: Observations, theory and models. *Reviews of Geophysics* 37:1-64.

Martin, P.S. 1984. Prehistoric overkill: the global model. Pp. 354-403 in P.S. Martin and R.G. Klein (eds.), *Quaternary Extinctions*. University of Arizona Press, Tucson.

Martinson, D.G., P.D Killworth, and A.L. Gordon. 1981. A convective model for the Weddell Polynya. *Journal of Physical Oceanography* 11(4):466-488.

Maslin, M.A., and S.J. Burns. 2000. Reconstruction of the Amazon basin effective moisture availability over the past 14,000 years. *Science* 290:2285-2287.

Mathewes, R.W. 1993. Evidence for Younger Dryas age cooling on the North Pacific coast of North America. *Quaternary Science Reviews* 12:321-332.

Mayewski, P.A., L.D. Meeker, M.S. Twickler, S. Whitlow, Q.Z. Yang, W.B. Lyons, and M. Prentice. 1997. Major features and forcing of high-latitude northern hemisphere atmospheric circulation using a 110,000-year-long glaciochemical series. *Journal of Geophysical Research* 102:26,345-26,366.

Mayle, F.E., A.J. Levesque, and L.C. Cwynar. 1993. Accelerator mass spectrometry ages for the Younger Dryas event in maritime eastern Canada. *Quaternary Research* 39:355-360.

McCreary J.P., and P. Lu. 1994. Interaction between the subtropical and equatorial ocean circulations—The subtropical cell. *Journal of Physical Oceanography* 24(2):466-497.

McGlone M.S., N.T. Moar, P. Wardle, and C.D. Meurk. 1997. Late-glacial and Holocene vegetation and environment of Campbell Island, far southern New Zealand. *Holocene* 7(1):1-12.

McLaren, A.S., R.G. Barry, and R.H.Bourke. 1990. Could Arctic ice be thinning? *Nature* 345:762.

McManus, J.F., R.F. Anderson, W.S. Broecker, M.O. Fleisher, and S.M. Higgins. 1998. Radiometrically determined sedimentary fluxes in the sub-polar North Atlantic during the last 140,000 years. *Earth and Planetary Science Letters* 155:29-43.

McPhaden, M.J., A.J. Busalacchi, R. Cheney, J.-R. Donguy, K.S. Gage, D. Halpern, M. Ji, P. Julian, G. Meyers, G.T. Mitchum, P.P. Niiler, J. Picaut, R.W. Reynolds, N. Smith, and K. Takeuchi. 1998. The Tropical Ocean-Global Atmosphere observing system: A decade of progress. *Journal of Geophysical Research* 103(14):169-240.

Meese, D.A., A.J. Gow, R.B. Alley, G.A. Zielinski, P.M. Grootes, M. Ram, K.C. Taylor, P.A. Mayewski, and J.F. Bolzan. 1997. The Greenland Ice Sheet Project 2 depth-age scale: methods and results. *Journal of Geophysical Research* 102:411-423.

Mellor P.S., and J. Boorman. 1995. The transmission and geographical spread of African horse sickness and bluetongue viruses. *Annals of Tropical Medicine and Parasitology* 89(1):1-15.

Meltzer, D. J., and J. I. Mead. 1983. The timing of Late Pleistocene mammalian extinctions in North America. *Quaternary Research* 19(1):130-135.

Mendelsohn R., W.D. Nordhaus, and D. Shaw. 1994. The impact of global warming on agriculture—A Ricardian analysis. *American Economic Review* 84(4):753-771.

Mendelsohn R. 2001. Abrupt climate change and forests. Paper prepared for Workshop on Economic and Ecological Impacts of Abrupt Climate Change. March 22-23, 2001.

Mercer, J.H. 1969. The Allerod oscillation: A European climate anomaly? *Arctic and Alpine Research* 1:227-234.

Mikolajewicz, U., and E. Maier-Reimer. 1994. Mixed boundary conditions in ocean general circulation models and their influence on the stability of the model's conveyor belt. *Journal of Geophysical Research* 99:22,633-22,644.

Mikolajewicz, U., and R. Voss. 2000. The role of the individual air-sea flux components in CO_2-induced changes of the ocean's circulation and climate. *Climate Dynamics* 16:627-642.

Mikolajewicz, U., T.J. Crowley, A. Schiller, and R. Voss. 1997. Modeling teleconnections between the North Atlantic and North Pacific during the Younger Dryas. *Nature* 387:384-387.

Minobe, S. 1997. Resonance in bidecadal and pentadecadal climate oscillations over the North Pacific: Role in climatic regime shifts. *Geophysical Research Letters* 26:855-858.

Minobe, S., and N. Mantua. 1999. Interdecadal modulation of interannual atmospheric and oceanic variability over the North Pacific. *Progress in Oceanography* 43(2-4):163-192.

Mitsch, W.J., and J.G. Gosselink. 2000. *Wetlands*. John Wiley and Sons, New York.

Molinari, R.L., R.A. Fine, W.D. Wilson, R.G. Curry, J. Abell, and M.S. McCartney. 1998. The arrival of recently formed Labrador Sea Water in the Deep Western Boundary Current at 26.5°N. *Geophysical Research Letters* 25:2249-2252.

Monahan, A.H., J.C. Fyfe, and G.M. Flato. 2000. A regime view of northern hemisphere atmospheric variability and change under global warming. *Geophysical Research Letters* 7:1139-1142.

Montoya, M., T.J. Crowley, and H. von Storch. 1998. Temperatures at the last interglacial simulated by a coupled ocean-atmosphere climate model. *Paleoceanography* 12(2):170-177.

Moreno, P.I., G.L. Jacobson, T.V. Lowell, and G.H. Denton. 2001. Interhemispheric climate links revealed by a late-glacial cooling episode in southern Chile. *Nature* 409(6822):804-808.

Morison, J.H., M. Steele, and R. Andersen. 1998. Hydrography of the upper Arctic Ocean measured from the nuclear submarine USS Pargo. *Deep-Sea Research* 45:15-38.

Morrill, C., J.T. Overpeck, and J.E. Cole. In review. A synthesis of abrupt changes in the Asian summer monsoon since the last deglaciation. *The Holocene.*

Mott, R.J. 1994. Wisconsinian late-glacial environmental change in Nova Scotia: A regional synthesis. *Journal of Quaternary Science* 9:155-160.

Muhs, D.R., and V.T. Holliday. 1995. Evidence of active dune sand on the Great Plains in the 19th century from accounts of early explorers. *Quaternary Research* 43:198-208.

Muhs, D.R., and V.T. Holliday. 2001. Origin of late quaternary dune fields on the Southern High Plains of Texas and New Mexico. *Geological Society of America Bulletin* 113:75-87.

Munk, W. 1966. Abyssal recipes. *Deep-Sea Research* 13:707-730.

Munk, W., and C. Wunsch. 1998. Abyssal recipes II, energetics of tidal and wind mixing. *Deep-Sea Research* 45:1977-2010.

Namias, J. 1960. Factors in the initiation, perpetuation and termination of drought. Extract of Publ. No. 51, International Association of Hydrological Sciences (IAHS) Commission of Surface Waters. [Available from IAHS Press, Institute of Hydrology, Wallingford, Oxfordshire OX10 8BB, UK.]

National Assessment Synthesis Team. 2000. *Climate Change Impacts on the United States: The Potential Consequences of Climate Variability and Change.* [Online]. Available: http://www.nacc.usgcrp.gov/ [2001, April 25].

National Research Council (NRC). 1992. *Water Transfers in the West: Efficiency, Equity, and the Environment.* National Academy Press, Washington, D.C.

National Research Council (NRC). 1998. *Decade-to-Century-Scale Climate Variability and Change: A Science Strategy.* National Academy Press, Washington, D.C.

National Research Council (NRC). 1999a. *From Monsoons to Microbes: Understanding the Ocean's Role in Human Health.* National Academy Press, Washington, D.C.

National Research Council (NRC). 1999b. *Making Climate Forecasts Matter.* National Academy Press, Washington, D.C.

National Research Council (NRC). 1999c. *Global Environmental Change: Research Pathways for the Next Decade.* National Academy Press, Washington, D.C.

National Research Council (NRC). 1999d. *Adequacy of Climate Observing Systems.* National Academy Press, Washington, D.C.

National Research Council (NRC). 2000. *Global Change Ecosystems Research.* National Academy Press, Washington, D.C.

National Research Council (NRC). 2001. *Climate Change Science: An Analysis of Some Key Questions.* National Academy Press, Washington, D.C.

Newnham, R.M., and D.J. Lowe. 2000. New Zealand paleoclimate. *Geology* 28:759-762.

Nicholson, S.E., C.J. Tucker, and M.B. Ba. 1998. Desertification, drought, and surface vegetation: An example from the West African Sahel. *Bulletin of the American Meteorological Society* 80:815-829.

Nicholson, S.E. 2001. Climatic and environmental change in Africa during the last two centuries. *Climate Research* Special Issue 17(2):123-144.

Nigam, S., I.M. Held, and S.W. Lyons. 1988. Linear Simulation of the Stationary Eddies in a Gcm.2. The mountain model. *Journal of Atmospheric Sciences* 45(9):1433-1452.

Nitsche, G., J.M. Wallace, and C. Kooperberg. 1994. Is there evidence of multiple equilibria in planetary wave amplitude statistics? *Journal of the Atmospheric Sciences* 51:314-322.

Nitta T., and S. Yamada. 1989. Recent warming of tropical sea surface temperature and its relationship to the Northern Hemisphere circulation. *Journal of the Meteorological Society of Japan* 67:187-193.

Nordhaus, W.D. 1994. *Managing the Global Commons: The Economics of Climate Change.* Massachusetts Institute of Technology Press, Cambridge.

Nordhaus, W.D. 2000. *The Economic Impacts of Abrupt Climatic Change.* Prepared for the Yale/NBER Workshop on the Societal Impact of Abrupt Climate Change, Snowmass, Colorado, July 24-25.

Nordhaus, W.D., and E. Kokkelenberg (eds.). 1999. *Nature's Numbers: Expanding the National Economic Accounts to Include the Environment.* Panel on Integrated Environmental and Economic Accounting, National Academy Press, Washington, D.C.

Nordhaus, W.D., and J. Boyer. 2000. *Warming the World: Economic Modeling of Global Warming.* Massachusetts Institute of Technology Press, Cambridge.

Norris, R.D., and U. Röhl. 1999. Carbon cycling and chronology of climate warming during the Palaeocene/Eocene transition. *Nature* 401:775-778.

O'Brien, S.R., P.A. Mayewski, L.D. Meeker, D.A. Meese, M.S. Twickler, and S.I. Whitlow. 1995. Complexity of Holocene climate as reconstructed from a Greenland ice core. *Science* 270:1962-1964.

Oeschger, H., J. Beer, U. Siegenthaler, B. Stauffer, W. Dansgaard, and C.C. Langway. 1984. Late glacial climate history from ice cores. Pp. 299-306 in J. Hansen and T. Takahashi (eds.), *Climate Processes and Climate Sensitivity*. American Geophysical Union, Washington, D.C.

Olsen, S.L.K. 1989. P 50 in D. Western and M.C. Pearl (eds.), *Conservation for the Twenty-first Century*. Oxford Univ. Press, UK.

Oppenheimer, M. 1998. Global warming and the stability of the West Antarctic Ice Sheet. *Nature* 393:325-332.

Oppo, D.W., and S.J. Lehman. 1995. Suborbital timescale variability of North Atlantic deep water formation during the last 200,000 years. *Paleoceanography* 12:191-205.

Oppo, D.W., M. Horowitz, and S.J. Lehman. 1997. Marine core evidence for reduced deepwater production during Termination II followed by a relatively stable substage 5e (Eemian). *Paleoceanography* 12(1):51-63.

Opsteegh, J.D., R.J. Haarsma, F.M. Selten, and A. Kattenberg. 1998. ECBILT: A dynamic alternative to mixed boundary conditions in ocean models. *Tellus* 50(A):348-367.

Orsi, A.H., G.C. Johnson, and J.L. Bullister. 1999. Circulation, mixing, and production of Antarctic Bottom Water. *Progress in Oceanography* 43:55-109.

Osborne, G., C. Clapperton, P.T. Davis, M. Reasoner, D.T. Rodbell, G.O. Seltzer, and G. Zielinski. 1995. Potential glacial evidence for the Younger Dryas event in the cordillera of North and South America. *Quaternary Science Reviews* 14(9):823-832.

Otterman, J. 1981. Satellite and field studies of man's impact on the surface in arid regions. *Tellus* 33:68-77.

Otto-Bliesner, B.L. 1999. El Niño, La Niña, and Sahel precipitation during the middle Holocene. *Geophysical Research Letters* 26:87-90.

Overpeck, J. 1996. Warm climate surprises. *Science* 271:1820-1821.

Overpeck, J., D. Rind, A. Lacis, and R. Healy. 1996. Possible role of dust-induced regional warming in abrupt climate change during the last glacial period. *Nature* 384:447-449.

Overpeck, J., K. Hughen, D. Hardy, R. Bradley, R. Case, M. Douglas, B. Finney, K. Gajewski, G. Jacoby, A. Jennings, S. Lamoreux, A. Lasca, G. MacDonald, J. Moore, M. Retelle, S. Smith, A. Wolfe, and G. Zielinski. 1997. Arctic environmental change of the last four centuries. *Science* 278:1251-1256.

Overpeck, J., and T. Webb. 2000. Nonglacial rapid events: Past and future. *Proceedings of the National Academy of Sciences of the United States of America* 97:1335-1338.

Palmer, W.M. 1965. Meteorological Drought. Research Pape No. 45, U.S. Weather Bureau, Washington, D.C. 58 pp. [Available from NOAA Library and Information Services Division, Washington, DC 20852.]

Palmer, T.N. 1993. Extended-range atmospheric prediction and the Lorenz model. *Bulletin of the American Meteorological Society* 74(1):49-65.

Palmer, T.N. 1999. A nonlinear dynamical perspective on climate prediction. *Journal of Climate* 12:575-591.

Parkinson, C.L. 2000. Variability of Arctic sea ice: The view from space, an 18-year record. *Arctic* 53(4):341-358.

Parson, E.A., and K. Fisher-Vanden. 1997. Integrated assessment of global climate change. *Annual Review of Energy and the Environment* 22:589-628.

Pascual, M., X. Rodo, S.P. Ellner, R. Colwell, and M.J. Bouma. 2000. Cholera dynamics and El Niño-Southern Oscillation. *Science* 289(5485):1766-1769.

Patrick, A., and R.C. Thunell. 1997. Tropical Pacific sea surface temperatures and upper water column thermal structure during the last glacial maximum. *Paleoceanography* 12: 649-657.

Patten, R.S., and D.H. Knight. 1994. Snow avalanches and vegetation patterns in Cascade Canyon, Grand Teton National Park, Wyoming, USA. *Arctic and Alpine Research* 26:35-41.

Pearce, F. 2000. Sink or swim. *New Scientist* 167:16-17.

Pearson, P.N., P.W. Ditchfield, J. Singano, K.G. Harcourt-Brown, C.J. Nicholas, R.K. Olsson, N.J. Shackleton, and M.A. Hall. 2001. Warm tropical sea surface temperatures in the late Cretaceous and Eocene epochs. *Nature* 413 (6855):481-487.

Perlwitz, J., and H.F. Graf. 1995. The statistical connection between tropospheric and stratospheric circulation of the Northern Hemisphere in winter. *Journal of Climate* 8:2281-2295.

Peteet, D.M. 1995. Global Younger Dryas? *Quaternary International* 28: 93-104.

Peteet, D.M. 2000. Sensitivity and rapidity of vegetational response to abrupt climate change. *Proceedings of the National Academy of Sciences of the United States of America* 97(4):1359-1361.

Peteet, D.M., and D.H. Mann. 1994. Late-glacial vegetational, tephra, and climatic history of southwestern Kodiak Island, Alaska. *Ecoscience* 1(3):255-267.

Peteet, D.M., J.S. Vogel, D.E. Nelson, J.R. Southon, J.R. Nickman, and L.E. Heusser. 1990. Younger Dryas climatic reversal in northeastern USA: AMS ages for an old problem. *Quaternary Research* 33:219-230.

Peteet, D.M., R.A. Daniels, L.E. Heusser, J.S. Vogel, J.R. Southon, and D.E. Nelson. 1993. Late-glacial pollen, macrofossils, and fish remains in northeastern USA-the Younger Dryas oscillation. *Quaternary Science Reviews* 12:597-612.

Peteet, D.M., A. Del Genio, and K.K. Lo. 1997. Sensitivity of Northern Hemisphere air temperatures and snow expansion to North Pacific sea surface temperatures in the GISS GCM. *Journal of Geophysical Research* 102(D20):23,781-23,791.

Peterson, L.C., G.H. Haug, K.A. Hughen, and U. Rohl. 2000. Rapid changes in the hydrologic cycle of the tropical Atlantic during the last glacial. *Science* 290:1947-1950.

Petit, J.R., J. Jouzel, D. Raynaud, N.I. Barkov, J.M. Barnola, I. Basile, M. Bender, J. Chappellaz, M. Davis, G. Delaygue, M. Delmotte, V.M. Kotlyakov, M. Legrand, V.Y. Lipenkov, C. Lorius, L. Pepin, C. Ritz, E. Saltzman, and M. Stievenard. 1999. Climate and atmospheric history of the past 420,000 years from the Vostok ice core, Antarctica. *Nature* 399:429-436.

Petoukhov, V., A. Ganopolski, V. Brovkin, M. Claussen, A. Eliseev, C. Kubatzki, and S. Rahmstorf. 2000. CLIMBER-2: A climate system model of intermediate complexity. Part I: model description and performance for present climate. *Climate Dynamics* 16:1-17.

Pickett, S.T., and P.S. White (eds.). 1985. *The Ecology of Natural Disturbance and Patch Dynamics*. Academic Press, San Diego.

Pickup, G. 1998. Desertification and climate change – the Australian perspective. *Climate Research* 11:51-63.

Pielke, Jr., R.A. 1999. Who decides forecasts and responsibilities in the 1997 Red River flood? *American Behavioral Science Review* 7(2):83-101.

Pielke, Jr., R.A., and C.W. Landsea. 1998. La Niña, El Niño, and Atlantic hurricane damages in the United States. *Bulletin of the American Meteorological Society* 80:2027-8033.

Pielke, Jr., R.A., and M.W. Downton. 2000. Precipitation and damaging floods: Trends in the United States, 1932-1997. *Journal of Climate* 13(20):3625-3637.

Pierrehumbert, R.T. 1999. Subtropical water vapor as a mediator of rapid global climate change. P. 394 in P.U. Clark, R.S. Webb, and L.D. Keigwin (eds.), *Mechanisms of Global Change at Millennial Time Scales*. Geophysical Monograph Series 112. American Geophysical Union, Washington, D.C.

Pierrehumbert, R.T. 2000. Climate change and the tropical Pacific: The sleeping dragon wakes. *Proceedings of the National Academy of Sciences of the United States of America* 97:1355-1358.

Pierrehumbert R.T., and R. Roca. 1998. Evidence for control of Atlantic subtropical humidity by large scale advection. *Geophysical Research Letters* 25(24):4537-4540.

Pimentel, D., and N. Kounang. 1998. Ecology of soil erosion in ecosystems. *Ecosystems* 1(5):416-426.

Piperno, D.R., M.B. Bush, and P.A. Colinvaux. 1990. Paleoenvironments and human settlements in late-glacial Panama. *Quaternary Research* 33:108-116.

Pisias, N.G., and A.C. Mix. 1997. Spatial and temporal oceanographic variability of the eastern equatorial Pacific during the later Pleistocene: Evidence from radiolaria microfossils. *Paleoceanography* 12:381-93.

Pollard, D., and S.L. Thompson. 1997. Climate and ice-sheet mass balance at the last glacial maximum from the GENESIS Version 2 global climate model. *Quaternary Science Reviews* 16:841-864.

Postel, S. 2000. Entering an era of water scarcity: The challenges ahead. *Ecological Applications* 10:941-948.

Poulsen, C.J., R.T. Pierrehumbert, and R.L. Jacob. 2001. Impact of ocean dynamics on the simulation of the Neoproterozoic "snowball earth." *Geophysical Research Letters* 28(8):1575-1578.

Precht, H., J. Christophersen, H. Hensel, and W. Larcher. 1973. *Temperature and Life*. Springer-Verlag, New York.

Radersma, S., and N. de Reider. 1996. Computed evapotranspiration of annual and perennial crops at different temporal and spatial scales using published parameter values. *Agricultural Water Management* 31:17-34.

Rahmstorf, S. 1994. Rapid climate transitions in a coupled ocean-atmosphere model. *Nature* 372:82-85.

Rahmstorf, S. 1995. Bifurcations of the Atlantic thermohaline circulation in response to changes in the hydrological cycle. *Nature* 378:145-149.

Rahmstorf, S. 1996. On the freshwater forcing and transport of the Atlantic thermohaline circulation. *Climate Dynamics* 12:799-811.

Rahmstorf, S., and J. Willebrand. 1995. The role of temperature feedback in stabilizing the thermohaline circulation. *Journal of Physical Oceanography* 25:787-805.

Rajagopalan, B., U. Lall, and M.A. Cane. 1999. Comment on "Reply to the comments of Trenberth and Hurrell." *American Meteorological Society* 80(12):2724-2726.

Reale, O., and P. Dirmeyer. 2000. Modeling the effects of vegetation on Mediterranean climate during the Roman Classical Period. Part I: Climate history and model sensitivity. *Global and Planetary Change* 25(3-4):163-184.

Reid, J.L. 1994. On the total geostrophic circulation of the North Atlantic Ocean: Flow patterns, tracers and transports. *Progress in Oceanography* 33:1-92.

Reid, J.L. 1998. On the total geostrophic circulation of the Pacific Ocean. Flow patterns, tracers and transports. *Progress in Oceanography* 39:263-352.

Reid, J.L. 2001. On the total geostrophic circulation of the Indian Ocean. Flow patterns, tracers and transports. *Progress in Oceanography* 39(4):263-352.

Reilly, J. Submitted. Abrupt climate change and agriculture. *Climatic Change*.

Reilly, J. ,and N. Hohmann. 1993. Climate change and agriculture: The role of international trade. *American Economic Association Papers and Proceedings* 83:306-312.

Reilly, J., N. Hohmann, and S. Kane. 1993. Climate change and agriculture: Global and regional effects using an economic model of international trade. *MIT-CEEPR 93-012WP.* Center for Energy and Environmental Policy Research, MIT, August 1993.

Reilly, J., and D. Schimmelpfennig. 2000. Irreversibility, uncertainty, and learning: Portraits of adaptation to long-term climate change. *Climatic Change* 45:253-278.

Reiter, P. 1998. Global-warming and vector-borne disease in temperate regions and at high altitude. *Lancet* 351(9105):839-840.

Renssen, H., and R.F.B. Isarin. 1998. Surface temperature in northwest Europe during the Younger Dryas: AGCM simulation compared with temperature reconstructions. *Climate Dynamics* 14:33-44.

Renssen, H., H. Goosse, T. Fichefet, and J.M. Campin. 2001. The 8.2 kyr BP event simulated by a global atmosphere-sea ice-ocean model. *Geophysical Research Letters* 28(8):1567-1570.

Renwick, J.A., and M.J. Revell. 1999. Blocking over the South Pacific and Rossby Wave Propagation. *Monthly Weather Review* 127(10):2233-2247.

Riebsame, W.E., S.A. Changnon, and T.R. Karl. 1991. *Drought and Natural Resources Management in the United States: Impacts and Implications of the 1987-89 Drought.* Westview Press. 174 pp.

Rigor, I.G., J.M. Wallace, and R.L. Colony. Submitted. On the response of sea ice to the Arctic Oscillation. *Journal of Climate.*

Rind, D. 1982. The influence of ground moisture conditions in North America on summer climate as modeled in the GISS GCM. *Monthly Weather Review* 110:1488-1494.

Rind, D., R. Goldberg, and R. Ruedy. 1989. Change in climate variability in the 21st century. *Climatic Change* 14:5-37.

Robert, C., and H. Maillot. 1990. Paleoenvironments in the Weddell Sea area and Antarctic climates, as deduced from clay mineral associations and geochemical data, ODP Leg 113. In P.F. Barker, J.P. Kennett, S. O'Connell, S. Berkowitz, W. Bryant, L. Burckle, P.K. Egeberg, D.K. Fuetterer, R. Gersonde, X. Golovchenko, N. Hamilton, L. Lawver, D.B. Lazarus, M.J. Lonsdale, B. Mohr, T. Nagao, C.P.G. Pereira, C. Pudsey, C.M. Robert, E.S. Schandl, V. Spiess, L.D. Stott, E. Thomas, K. Thompson, S.W. Wise, D.M. Kennett, A. Masterson, and N.J. Stewart (eds.), *Proceedings of the Ocean Drilling Program: Scientific Results,* Vol. 113. Ocean Drilling Program, College Station, Texas.

Rodbell, D.T., and G.O. Seltzer. 2000. Rapid ice margin fluctuations during the Younger Dryas in the tropical Andes. *Quaternary Research* 54(3):328-338.

Rodbell, D.T., G.O. Seltzer, D.M. Anderson, M.B Abbott, D.B. Enfield, and J.H. Newman. 1999. An ~15,000 record of El Niño-driven alluviation in southwestern Ecuador. *Science* 283:516-520.

Rodwell, M.J., and B.J. Hoskins. 2001. Subtropical anticyclones and summer monsoons. *Journal of Climate* 14:3192-3211.

Rodwell, M.J., D.P. Rowell, and C.K. Folland. 1999. Oceanic forcing of the wintertime North Atlantic Oscillation and European climate. *Nature* 398:320-323.

Roe, G.H., and R.S. Lindzen. 2001. The mutual interaction between continental-scale ice sheets and atmospheric stationary waves. *Journal of Climate* 14:1450-1465.

Rooth, C. 1982. Hydrology and ocean circulation. *Progress in Oceanography* 11:131-149.

Rothrock, D.A., Y. Yu, and G.A. Maykut, 1999. Thinning of the Arctic sea-ice cover. *Geophysical Research Letters* 26:3469-3472.

Rosenberg, N.J. 1978. *North American Drought.* American Association for the Advancement of Science, Westview Press, New York.

Rosenzweig, C., and M. L. Parry. 1994. Potential impact of climate change on world food supply. *Nature* 367:133-138.

Rothrock, D.A., Y. Yu, and G.A. Maykut. 1999. Thinning of the Arctic sea-ice cover. *Geophysical Research Letters* 26:3469-3472.

Roy, D.B., and T.H. Sparks. 2000. Phenology of British butterflies and climate change. *Global Change Biology* 6:407-416.

Ruddiman, W.F., and A. McIntyre. 1981. The mode and mechanism of the last deglaciation: Oceanic evidence. *Quaternary Research* 16:125-134.

Sachs J.P., and S.J. Lehman. 1999. Subtropical North Atlantic temperatures 60,000 to 30,000 years ago. *Science* 286:756-759.

Sampat, P. 2000. Deep trouble: The hidden threat of groundwater pollution. *Worldwatch Paper #154*. Worldwatch Institute, Washington, D.C. 55 pp.

Sandweiss, D.H., K.A. Maasch, R.L. Burger, J.B. Richardson III, H.B. Rollins, and A. Clement. 2001. Variation in Holocene El Niño frequencies: Climate records and cultural consequences in ancient Peru. *Geology* 29(7):603-606.

Sarnthein, M., K. Winn, S.J.A. Jung, J.C. Duplessy, L. Labeyrie, H. Erlenkeuser, and G. Ganssen. 1994. Changes in east Atlantic deepwater circulation over the last 30,000 years: Eight time slice reconstructions. *Paleoceanography* 9:209-267.

Scherer, R.P., A. Aldahan, S. Tulaczyk, G. Possnert, H. Engelhardt, and B. Kamb. 1998. Pleistocene collapse of the West Antarctic ice sheet. *Science* 281:82-85.

Schiller, A., U. Mikolajewicz, and R. Voss. 1997. The stability of the North Atlantic thermohaline circulation in a coupled ocean-atmosphere general circulation model. *Climate Dynamics* 13:325-347.

Schlesinger, W.H. 1991. *Biogeochemistry: An Analysis of Global Change*. Academic Press, San Diego, CA. 443 pp.

Schlosser, P., B. Kromer, R. Weppernig, H.H. Loosli, R. Bayer, G. Bonani, and M. Suter. 1994. The distribution of C-14 and Ar-39 in the Weddell Sea. *Journal of Geophysical Research* 99(C5):10,275-10,287.

Schmitt, R.W., P.S. Bogden, and C.E. Dorman. 1989. Evaporation minus precipitation and density fluxes for the North Atlantic. *Journal of Physical Oceanography* 19:1208-1221.

Schmittner, A. and A.J. Weaver. 2001. Dependence of multiple climate states on ocean mixing parameters. *Geophysical Research Letters* 28:1027-1030.

Schmittner, A., C. Appenzeller, and T.F. Stocker. 2000. Enhanced Atlantic freshwater export during El Niño. *Geophysical Research Letters* 27:1163-1166.

Schmitz, B., V. Pujalte, and K. Nunez-Betelu. 2001. Climate and sea-level perturbations during the initial Eocene thermal maximum; evidence from siliciclastic units in the Basque Basin (Ermua, Zumaia and Trabakua Pass), northern Spain. *Palaeogeography, Palaeoclimatology, Palaeoecology* 165:299-320.

Schulz, H., U.V. Rad, and H. Erlenkeuser. 1998. Correlation between Arabian Sea and Greenland climate oscillations of the past 110,000 years. *Nature* 393:54-57.

Schwartz, M.D., and B.E. Reiter. 2000. Changes in North American spring. *International Journal of Climatology* 20:929-932.

Scott, L., M. Steenhamp, and P.B. Beaumont. 1995. Paleoenvironmental conditions in South Africa at the Pleistocene-Holocene transition. *Quaternary Science Reviews* 14:937-948.

Scott, J.R., J. Marotzke, and P.H. Stone. 1999. Interhemispheric thermohaline circulation in a coupled box model. *Journal of Physical Oceanography* 29:351-365.

Seager R., A.C. Clement, and M.A. Cane. 2000. Glacial cooling in the tropics: Exploring the roles of tropospheric water vapor, surface wind speed, and boundary layer processes. *Journal of the Atmospheric Sciences* 57(13):2144-2157.

Seager R., D.S. Battisti, J. Yin, N. Gordon, N. Naik, A.C. Clement, and M.A. Cand. Submitted. Is the Gulf Stream responsible for Europe's mild winters? *Quarterly Journal of the Royal Meteorological Society*.

Sellers, W.D. 1969. A global climate model based on the energy balance of the earth-atmosphere system. *Journal of Applied Meteorology* 8:392-400.

Sellinger, C.E., and F. Quinn. 1999. *Proceedings of the Great Lakes Paleo-Levels Workshop: The Last 4000 Years*. NOAA Technical Memorandum ERL GLERL-113. National Oceanic and Atmosphericc Admnistration, Silver Spring, MD. 43 pp.

Severinghaus, J.P., T. Sowers, E.J. Brook, R.B. Alley, and M.L. Bender. 1998. Timing of abrupt climate change at the end of the Younger Dryas interval from thermally fractionated gases in polar ice. *Nature* 391:141-146.

Shafer, S.L., P.J. Bartlein, and R.S. Thompson. 2001. Potential changes in the distributions of western North America tree and shrub taxa under future climate scenarios. *Ecosystems* 4(3):200-215.

Shane, L.C.K., and K.H. Anderson. 1993. Intensity, gradients, and reversals in late-glacial environmental change in east-central North America. *Quaternary Science Reviews* 12:307-320.

Shindell, D.T., R.L. Miller, G. Schmidt, and L. Pandolfo. 1999. Simulation of recent northern winter climate trends by greenhouse-gas forcing. *Nature* 399:452-455.

Shindell, D.T., G.A. Schmidt, R.L. Miller, and D. Rind. 2001. Northern Hemisphere winter climate response to greenhouse gas, ozone, solar, and volcanic forcing. *Journal of Geophysical Research* 106:7193-7210.

Shukla, H., and Y. Mintz. 1982. The influence of land surface evapotranspiration on the earth's climate. *Science* 215:1498-1501.

Sillett T.S., R.T. Holmes, and T.W. Sherry. 2000. Impacts of a global climate cycle on population dynamics of a migratory songbird. *Science* 288(5473):2040-2042.

Singer, C., J. Shulmeister, and B. McLea. 1998. Evidence against a significant Younger Dryas cooling event in New Zealand. *Science* 281(5378):812-814.

Sissons, J.B. 1967. *The Evolution of Scotland's Scenery*. Oliver and Boyd, Edinburgh.

Skaggs, R.H. 1975. Drought in the United States 1931-1940. *Annual Association of American Geography* 65:391-402.

Smethie, W.M., and R.A. Fine. 2001. Rates of North Atlantic deep water formation calculated from chlorofluorocarbon inventories. *Deep-Sea Research* 48:189-215.

Smethie, W.M., R.A. Fine, A. Putzka, and E.P. Jones. 2000. Tracing the flow of North Atlantic Deep Water using chlorofluorocarbons. *Journal of Geophysical Research* 105(C6):14,297-14,323.

Smith, J., M.J. Risk, H.P. Schwarcz, and T.A. McConaugher. 1997. Rapid climate change in the North Atlantic during the Younger Dryas recorded by deep-sea corals. *Nature* 386:818-820.

Sohngen, B., and R. Mendelsohn. 1998. Valuing the market impact of large-scale ecological change: The effect of climate change on U.S. timber. *American Economic Review* 88:686-710.

Somayao, C.R., E.T. Kanemasu, and T.W. Brakke. 1980. Using leaf temperature to assess evapotranspiration and advection. *Agricultural Meteorology* 22:153-166.

Sparks, T., H. Heyen, O. Braslavska, and E. Lehikoinen. 1999. Are European birds migrating earlier? *BTO News* 223:8-9.

Spear, R.W., M.B. Davis, and L.C.K. Shane. 1994. Late Quaternary history of low and mid-elevation vegetation in the White Mountains of New Hampshire. *Ecological Monographs* 64:85-109.

Speer J.H., T.W. Swetnam, B.E. Wickman, and A. Youngblood. 2001. Changes in pandora moth outbreak dynamics during the past 622 years. *Ecology* 82(3):679-697.

Spinelli, G. 1996. A statistical analysis of ice-accumulation level and variability in the GISP2 ice core and a reexamination of the age of the termination of the Younger Dryas cooling episode. *Earth System Science Center Technical Report No. 96-001.* The Pennsylvania State University, University Park.

Stager J.C., and P.A. Mayewski. 1997. Abrupt early to mid-Holocene climatic transition registered at the equator and the poles. 1997. *Science* 276(5320):1834-1836.

Stahle, D.W., E.R. Cook, M.K. Cleaveland, M.D. Therrell, D.M. Meko, H.D. Grissino-Mayer, E. Watson, and B.H. Luckman. 2000. Tree-ring data document 16th century mega-drought over North America. *EOS, Transactions, American Geophysical Union* 81(12): 121,125.

Stahle, D.W., M.K. Cleaveland, D.B. Blanton, M.D. Therrell, and D.A. Gay. 1998. The Lost Colony and Jamestown droughts. *Science* 280:564-566.

Stamm, J.F., E.F. Wood, and D.P. Lettenmaier. 1994. Sensitivity of a GCM simulation of global climate to the representation of land-surface hydrology. *Journal of Climate* 7:1218-1239.

Steig, E.J., E.J. Brook, J.W.C. White, C.M. Sucher, M.L. Bender, S.J. Lehman, D.L. Morse, E.D. Waddington and G.D. Clow. 1998. Synchronous climate changes in Antarctica and the North Atlantic. *Science* 282:92-95.

Stine, S. 1994. Extreme and persistent drought in California and Patagonia during mediaeval time. *Nature* 369:546-549.

Stocker, T.F. 1998. The seesaw effect. *Science* 282:61-62.

Stocker, T.F. 2000. Past and future reorganisations in the climate system. *Quaternary Science Reviews* 19:301-319.

Stocker, T.F., D.G. Wright, and W.S. Broecker. 1992. The influence of high-latitude surface forcing on the global thermohaline circulation. *Paleoceanography* 7:529-41.

Stocker, T.F., and A. Schmittner. 1997. Influence of CO_2 emission rates on the stability of the thermohaline circulation. *Nature* 388:862-865.

Stocker, T.F., and D.G. Wright. 1991. Rapid transitions of the ocean's deep circulation induced by changes in surface water fluxes. *Nature* 351:729-732.

Stocker, T.F., and O. Marchal. 2000. Abrupt climate change in the computer: is it real? *Proceedings of the National Academy of Sciences of the United States of America* 97:1362-1365.

Stocker, T.F., and O. Marchal. 2001. Recent progress in paleoclimate modeling: Climate models of reduced complexity. *PAGES Newsletter* 9(1):4-7.

Stocker, T.F., D.G. Wright, and L.A. Mysak. 1992. A zonally averaged, coupled ocean-atmosphere model for paleoclimate studies. *Journal of Climate* 5:773-797.

Stommel, H. 1961. Thermohaline convection with two stable regimes of flow. *Tellus* 13:224-230.

Stommel, H., and E. Stommel. 1983. *Volcano Weather: The Story of the Year Without a Summer.* Seven Seas Press, Newport, RI. 177 pp.

Stott, P.A., S.F.B. Tett, G.S. Jones, M.R. Allen, J.F.B. Mitchell, and G.J. Jenkins. 2000. External control of 20th Century temperature by natural and anthropogenic forcings. *Science* 290:2133-2137.

Street, D.G., and M.H. Glantz, 2000. Exploring the concept of climate surprise. *Global Environmental Change* 10:97-107.

Street-Perrott, F.A., and R.A. Perrott. 1990. Abrupt climate fluctuations in the tropics: The influence of Atlantic Ocean circulation. *Nature* 343:607-611.

Street-Perrott, F.A., J.A. Holmes, M.P. Waller, M.J. Allen, N.G.H. Barber, P.A. Fothergill, D.D. Harkness, M. Ivanovich, D. Kroon, and R.A. Ferrott. 2000. Drought and dust deposition in the West African Sahel: A 5500-year record from Kajemarum Oasis, northeastern Nigeria. *The Holocene* 10:293-302.

Stuiver, M., P.M. Grootes, and T.F. Brazunias. 1995. The GISP2 $\delta^{18}O$ record of the past 16,500 years and the role of the sun, ocean and volcanoes. *Quaternary Research* 44:341-354.

Sud, Y.C., and A. Molod. 1988. A GCM simulation study of the influence of Saharan evapotranspiration and surface-albedo anomalies on July circulation and rainfall. *Monthly Weather Review* 116:2388-2400.

Sutera, A. 1986. Probability density distribution of large-scale atmospheric flow. *Advances in Geophysics* 29:227-249.

Svensmark, H., and E. FriisChristensen. 1997. Variation of cosmic ray flux and global cloud coverage—A missing link in solar-climate relationships. *Journal of Atmospheric and Solar-Terrestrial Physics* 59(11):1225-1232.

Svensmark, H. 1998. Influence of cosmic rays on Earth's climate. *Physical Review Letters* 81(22):5027-5030.

Swetnam T.W., C.D. Allen CD, and J.L. Betancourt. 1999. Applied historical ecology: Using the past to manage for the future. *Ecological Applications* 9(4):1189-1206.

Swetnam, T.W., and J.L. Betancourt. 1990. Fire-southern oscillation relations in the southwestern United States. *Science* 249:1017-1020.

Swetnam, T.W., and J.L. Betancourt. 1998. Mesoscale disturbance and ecological response to decadal climatic variability in the American Southwest. *Journal of Climate* 11:3128-3147.

Sy, A., M. Rhein, J.R.N. Lazier, K.P. Koltermann, J. Meincke, A. Putzka, and M. Bersch. 1997. Surprisingly rapid spreading of newly formed intermediate waters across the North Atlantic ocean. *Nature* 386:675-679.

Talley, L.D. 1999. Some aspects of ocean heat transport by the shallow, intermediate and deep overturning circulations. Pp. 1-22 in P.U. Clark, R.S. Webb, and L.D. Keigwin (eds.), *Mechanisms of Global Climate Change at Millenial Time Scales*. Geophysical Monograph Series 112. American Geophysical Union, Washington, D.C.

Taylor, K.C., G.W. Lamorey, G.A. Doyle, R.B. Alley, P.M. Grootes, P.A. Mayewski, J.W.C. White, and L.K. Barlow. 1993. The "flickering switch" of late Pleistocene climate change. *Nature* 361:432-436.

Taylor, K.C., P.A. Mayewski, M.S. Twickler, and S.I. Whitlow. 1996. Biomass burning recorded in the GISP2 ice core: A record from eastern Canada. *Holocene* 6:1-6.

Taylor, K.C., P.A. Mayewski, R.B. Alley, E.J. Brook, A.J. Gow, P.M. Grootes, D.A. Meese, E.S. Saltzman, J.P. Severinghaus, M.S. Twickler, J.W.C. White, S. Whitlow, and G.A. Zielinski. 1997. The Holocene/Younger Dryas transition recorded at Summit, Greenland. *Science* 278:825-827.

Teller, J.T. 1990. Meltwater and precipitation runoff to the North Atlantic, Arctic, and Gulf of Mexico from the Laurentide ice sheet and adjacent regions during the Younger Dryas. *Paleoceanography* 5:897-905.

Teller, J.T. In press. Freshwater outbursts to the oceans from glacial Lake Agassiz and climate change during the last deglaciation. *Quaternary Science Reviews*.

Thomas, E., and N.J. Shackleton. 1996. The Paleocene-Eocene benthic foraminiferal extinction and stable isotope anomalies. Pp. 401-441 in R. Knox, R.M. Corfield, and R.E. Dunay (eds.), *Correlation of the Early Paleogene in Northwest Europe*. Special Publications of the Geological Society of London, Vol. 101.

Thompson, D.W.J., and J.M. Wallace. 1998. The Arctic Oscillation signature in the wintertime geopotential height and temperature fields. *Geophysical Research Letters* 25:1297-1300.

Thompson, D.W.J., J.M. Wallace, and G.C. Hegerl. 2000. Annular modes in the extratropical circulation. Part II: Trends. *Journal of Climate* 13:1018-1036.

Thompson, L.G., E. Mosley-Thompson, M.E. Davis, P.N. Lin, K.A. Henderson, J. Cole-Dai, J.F. Bolzan, and K.B. Liu. 1995. Late-glacial stage and Holocene tropical ice core records from Huascaran, Peru. *Science* 269:46-50.

Thompson, L.G., E. Mosley-Thompson, T.A. Sowers, K.A. Henderson, V.S. Zagorodnov, P.N. Lin, V.N. Mikhalenko, R.K. Campen, J.F. Bolzan, J. Cole-Dai, and B. Francou. 1998. A 25,000-year tropical climate history from Bolivian ice cores. *Science* 282:1858-1864.

Timmermann, A., and G. Lohmann. 2000. Noise-induced transitions in a simplified model of the thermohaline circulation. *Journal of Physical Oceanography* 30:1891-1900

Timmermann A., J. Oberhuber, A. Bacher, M. Esch, M. Latif, and E. Roeckner. 1999. Increased El Niño frequency in a climate model forced by future greenhouse warming. *Nature* 398(6729):694-697.

Tinner, W., and A.F. Lotter. 2001. Central European vegetation response to abrupt climate change at 8.2 ka. *Geology* 29(6):551-554.

Toggweiler, R., and B. Samuels. 1995. Effect of the Drake Passage on the global thermohaline circulation. *Deep-Sea Research* 42:477-500.

Toresen, R., and O.J. Ostvedt. 2000. Variation in abundance of Norwegian spring-spawned herring (*Clupea harengus*, Clupeidae) throughout the 20th Century and the influence of climatic fluctuations. *Fish and Fisheries* 1:231-256.

Toth, F. 2000. *Invoking Apocalypse: Coping with the Risk of Abrupt Climate Change.* Prepared for the Yale/NBER Workshop on the Societal Impact of Abrupt Climate Change, Snowmass, Colorado, July 24-25.

Toth, F., M. Mwandosya, C. Carraro, J. Christensen, J. Edmonds, B. Flannery, C. Gay-Garcia, H. Lee, K.M. Meyer-Abich, E. Nikitina, A. Rahman, Y. Ruqiu, A. Villavicencio, Y. Wake, and J. Weyant. 2001. Decision making frameworks. In *Climate Change 2001: Mitigation.* Intergovernmental Panel on Climate Change. Cambridge University Press, UK.

Trauth, M.H., A Deino, and M.R. Strecker. 2001. Response of the East African climate to orbital forcing during the last interglacial (130-117 ka) and the early last glacial (117-60 ka). *Geology* 29(6):499-502.

Trenberth, K.E. 1990. Recent observed interdecadal climate changes in the Northern Hemisphere. *Bulletin of the American Meteorological Society* 71:988-993.

Trenberth, K.E., and T.J. Hoar. 1996. The 1990-1995 El Niño-Southern Oscillation event: Longest on record. *Geophysical Research Letters* 23:57-60.

Trenberth, K.E. 1999. Atmospheric moisture recycling: Role of advection and local evaporation. *Journal of Climate* 12:1368-1381.

Trenberth K.E., and J.W. Hurrell. 1999a. Comments on "The interpretation of short climate records with comments on the North Atlantic and Southern Oscillations." *Bulletin of the American Meteorological Society* 80(12):2721-2722.

Trenberth K.E., and J.W. Hurrell. 1999b. Reply to Rajagopalan, Lall, and Cane's comment about "The interpretation of short climate records with comments on the North Atlantic and Southern Oscillations." *Bulletin of the American Meteorological Society* 80(12):2726-2728.

Tudhope, A.W., C.P. Chilcott, M.T. McCulloch, E.R. Cook, J. Chappell, R.M. Ellam, C.W. Lea, J.M. Lough, and G.B. Shimmield. 2001. Variability in the El Niño-Southern Oscillation through a glacial-interglacial cycle. *Science* 291:1511-1517.

Tziperman, E. 1997. Inherently unstable climate behavior due to weak thermohaline ocean circulation. *Nature* 386:592-595.

United Nations (UN). 1980. M.K. Biswas and A.K. Biswas (eds.), *Desertification.* Pergamon Press, London.

Urban, F.E., J.E. Cole, and J.T. Overpeck. 2000. Influence of mean climate change on climate variability from a 155-year tropical Pacific coral record. *Nature* 407:989-993.

van der Hammen, T., and H. Hooghiemstra. 1995. The El Abra stadial, a Younger Dryas equivalent in Colombia. *Quaternary Science Reviews* 14(9):841-851.

van Kolfschoten, T., and P.L. Gibbard. 2000. The Eemian—local sequences, global perspectives: Introduction. *Geologie en Mijnbouw* 79(2-3):129-133.

U.S. Army Corps of Engineers. 1994. *The Great Flood of 1993 Post-Flood Report: Upper Mississippi River and Lower Missouri River Basins–Main Report*. North Central Division. 609 pp.

Verschuren, D., K.R. Laird, and B.F. Cumming. 2000. Rainfall and drought in equatorial east Africa during the past 1,100 years. *Nature* 403:410-414.

Villalba R., and T.T. Veblen. 1998. Influences of large-scale climatic variability on episodic tree mortality in northern Patagonia. *Ecology* 79(8):2624-2640.

Vinnikov, K.Y., A. Robock, R.J. Stouffer, J.E. Walsh, C.L. Parkinson, D.J. Cavalieri, J.F.B. Mitchell, D. Garrett, and V.F. Zakharov. 1999. Global warming and Northern Hemisphere sea ice extent. *Science* 286:1934-1937.

Vitousek, P.M., P.R. Ehrlich, A.H. Ehrlich, and P.A. Matson. 1986. Human appropriation of the products of photosynthesis. *Bioscience* 36:368.

Vitousek, P.M., H.A. Mooney, J. Lubchenco, and J.M. Melillo. 1997. Human domination of earth's ecosystems. *Science* 277:494-499.

Volodin, E.M., and V.Y. Galin. 1999. Interpretation of winter warming on Northern Hemisphere continents in 1977-94. *Journal of Climate* 12 (10): 2947-2955 8:23-32.

von Grafenstein, U., H. Erlenkeuser, A. Brauer, J. Jouzel, and S.J. Johnsen. 1999. A Mid-European decadal isotope-climate record from 15,500 to 5000 years BP. *Science* 284(5420):1654-1657.

Wadhams, P., and N.R. Davis. 2000. Further evidence of ice thinning in the Arctic Ocean. *Geophysical Research Letters* 27:3973-3975.

Walker, M.J.C. 1995. Climatic changes in Europe during the last glacial/interglacial transition. *Quaternary International* 28:63-76.

Walker, M.J.C., G.R. Coope, and J.J. Lowe. 1993. The Devensian (Weichselian) late-glacial paleoenvironmental record from Gransmoor, East Yorkshire, England. *Quaternary Science Reviews* 12:659-680.

Walker, L.R., and M.R. Willig (eds.). 1999. *Ecology of Disturbed Ground*. Elsevier, Amsterdam.

Wallace, J.M. 2000. North Atlantic Oscillation/Northern Hemisphere annular mode: One phenomenon, two paradigms. *Quarterly Journal of the Royal Meteorological Society* 126:791-805.

Walsh, J.E., W.L. Chapman, and T.L. Shy. 1996. Recent decrease of sea level pressure in the central Arctic. *Journal of Climate* 9:480-486.

Walsh, J.E., W.H. Jasperson, and B. Ross. 1985. Influence of snow cover and soil moisture on monthly air temperatures. *Monthly Weather Reviews* 113:756-768.

Wang, X.L., P.H. Stone, and J. Marotzke. 1999. Global thermohaline circulation. Part I: Sensitivity to atmospheric moisture transport. *Journal of Climate* 12:71-82.

Wang, G.L., and E.A.B. Eltahir. 2000a. Biosphere-atmosphere interaction over west Africa. II. Multiple climate equilibria. *Quarterly Journal of the Royal Meteorological Society* 126:1261-1280.

Wang, G.L., and E.A.B. Eltahir. 2000b. Role of vegetation dynamics in enhancing the low-frequency variability of the Sahel rainfall. *Water Resources Research* 36:1013-1021.

Warren, B.A. 1983. Why is no deep water formed in the North Pacific? *Journal of Marine Research* 41:327-347.

Watts, W.A. 1980. Regional variation in the response of vegetation to Lateglacial climatic events in Europe. In J.J. Lowe, J.M. Gray, and J.E. Robinson (eds.), *Studies in the Lateglacial of Northwest Europe*. Pergamon Press, Oxford, UK.

Weertman, J. 1974. Stability of the junction of an ice sheet and an ice shelf. *Journal of Glaciology* 13:3-11.

Weidick A., H. Oerter, N. Reeh, H. Thomsen, and L. Thorning. 1990. The recession of the inland ice margin during the Holocene climatic optimum in the Jakobshavn-Isfjord area of west Greenland. *Global and Planetary Change* 82(3-4):389-399.

Weiss, H., and R.S. Bradley. 2001. What drives societal collapse? *Science* 291:609-610.

Weiss, H., M.A. Courty, W. Wetterstrom, L. Senior, R. Meadow, F. Guichard, and A. Curnow. 1993. The genesis and collapse of third millennium North Mesopotamian civilization. *Science* 261(20):995-1004.

Wendler, G., and F. Eaton. 1983. On the desertification in the Sahel zone. Part 1. Ground observations. *Climate Change* 5:365-380.

West, J.J., M.J. Small, and H. Dowlatabadi. 2001. Storms, investor decisions, and the economic impacts of sea level rise. *Climatic Change* 48(2-3):317-342.

Weyant, J.P., O. Davidson, H. Dowlatabadi, J. Edmonds, M. Grubb, E.A. Parson, R. Richels, J. Rotmans, P.R. Shukla, and R.S.J. Tol. 1996. Integrated assessment of climate change: An overview and comparison of approaches and results. Pp. 367-396 in *Climate Change 1995: Economic and Social Dimensions of Climate Change*. Intergovernmental Panel on Climate Change, Cambridge University Press, UK.

White, W.B., and R. Peterson. 1996. An Antarctic Circumpolar Wave in surface pressure, wind, temperature, and sea ice extent. *Nature* 380:699-702.

Whitworth, T., B.A. Warren, W.D. Nowlin, S.B. Rutz, R.D. Pillsbury, and M.I. Moore. 1999. On the deep western-boundary current in the Southwest Pacific Basin. *Progress in Oceanography* 43:1-54.

Winton, M. and E.S. Sarachik. 1993. Thermohaline oscillations induced by strong steady salinity forcing of ocean general-circulation models. *Journal of Physical Oceanography* 23:1389-1410.

Wong, A.P.S., N.L. Bindoff, and J.A. Church. 1999. Large-scale freshening of intermediate waters in the Pacific and Indian oceans. *Nature* 400:440-443.

Wood, R.A., A.B. Keen, J.F.B. Mitchell, and J.M. Gregory. 1999. Changing spatial structure of the thermohaline circulation in response to atmospheric CO_2 forcing in a climate model. *Nature* 399:572-575.

Woodhouse, C.A. ,and J.T. Overpeck. 1998. Two thousand years of drought variability in the central United States. *Bulletin of the American Meteorological Society* 79:2693-2714.

Woodward, F.I. 1987. *Climate and Plant Distribution*. Cambridge University Press, UK.

World Bank. 2001. *World Development Indicators, 2000*. World Bank, Washington, D.C.

Wright, D.G., and T.F. Stocker. 1991. A zonally averaged ocean model for the thermohaline circulation. Part I: Model development and flow dynamics. *Journal of Physical Oceanography* 21:1713-1724.

Xue, Y.K., and J. Shukla. 1993. The influence of land surface properties on Sahel climate: Part I. Desertification. *Journal of Climate* 6:2232-2245.

Xue, Y.K., M.J. Fennessy, and P.J. Sellers. 1996. The impact of vegetation properties on U.S. summer weather prediction. *Journal of Geophysical Research* 101:7419-7430.

Yang, R., M.J. Fennessy, and J. Shukla. 1994. The influence of initial soil wetness on medium-range surface weather forecasts. *Monthly Weather Reviews* 122:471-483.

Yeh, T.C., R.T. Wetherald, and S. Manabe. 1984. The effect of soil moisture on the short-term climate and hydrology change—a numerical experiment. *Monthly Weather Reviews* 112:474-490.

Yiou, P., K. Fuhrer, L.D. Meeker, J. Jouzel, S. Johnsen, and P.A. Mayewski. 1997. Paleoclimatic variability inferred from the spectral analysis of Greenland and Antarctic ice-core data. *Journal of Geophysical Research* 102:441-454.

Yohe, G.W., and M.E. Schlesinger. 1998. Sea-level change: the expected economic cost of protection or abandonment in the United States. *Climatic Change* 38:337-342.

Yu, Z., J.H. McAndrews, and U. Eicher. 1997. Middle Holocene dry climate caused by change in atmospheric circulation patterns: Evidence from lake levels and stable isotopes. *Geology* 25:251-254.

Yu, Z., and U. Eicher. 1998. Abrupt climate oscillations during the last deglaciation in central North America. *Science* 282:2235-2238.

Yueh, S.H., and R. Kwok. 1998. Arctic sea ice extent and melt onset from NSCAT observations. *Geophysical Research Letters* 25:4369-4372.

Zaucker, F., and W.S. Broecker. 1992. The influence of atmospheric moisture transport on the fresh water balance of the Atlantic drainage basin: General circulation model simulations and observations. *Journal of Geophysical Research* 97:2765-2773.

Zhang, Y., J.M. Wallace, and D.S. Battisti. 1997. ENSO-like interdecadal variability: 1900-93. *Journal of Climate* 10:1004-1020.

Zielinski, G.A., P.A. Mayewski, L.D. Meeker, K. Gronvold, M.S. Germani, S. Whitlow, M.S. Twickler, and K. Taylor. 1997. Volcanic aerosol records and tephrochronology of the Summit, Greenland, ice cores. *Journal of Geophysical Research* 102(C12):26,625-26,640.

Appendixes

A

Committee and Staff Biographies

Richard Alley (*Chair*) is Evan Pugh Professor of Geosciences and an Associate of the College of Earth and Mineral Sciences Environment Institute at the Pennsylvania State University. Professor Alley studies past climate change by analyzing ice cores from Greenland and Antarctica. He has demonstrated that exceptionally large climate changes have occurred in as little as a single year. His work on deformation of subglacial tills has given new insights to ice-sheet stability and the interpretation of glacial deposits. Ongoing work on ice-flow modeling may lead to predictions of future sea-level change. Related interests include metamorphic textures of ice, transformation of snow to ice, microwave remote sensing of ice, origins of ice stratification, controls on snowfall, monitoring of past storm tracks. Dr. Alley is a member of the Polar Research Board and a Fellow of the American Geophysical Union.

Jochem Marotzke is a Professor at the Southampton Oceanography Centre in the United Kingdom. His research interests are in large-scale ocean circulation and the role of the ocean in climate and abrupt climate change. He has worked on the stability of the ocean's thermohaline circulation, large-scale ocean-atmosphere interactions, and model-data syntheses applied to the Atlantic and Indian oceans. He then focused on the fluid dynamics of the thermohaline circulation, especially the role of oceanic mixing. More recently, he has explored the interactions of the thermohaline circulation with sea ice and the ocean carbon cycle, and also the role of the thermohaline circulation for paleoclimates.

William Nordhaus is the A. Whitney Griswold Professor of Economics at Yale University and was a member of the U.S. President's Council of Economic Advisers. His research focuses on economic growth and natural resources as well as the question of the extent to which resources constrain economic growth. Recently, his work has focused on the economics of global warming, including the construction of integrated economic and scientific models to determine an efficient path for coping with climate change. He is a member of the National Academy of Sciences, a member and senior advisor of the Brookings Panel on Economic Activity, and a Fellow of the American Academy of Arts and Sciences. Professor Nordhaus has served on the executive committees of the American Economic Association and the Eastern Economic Association. He has served as a member of the Board on Sustainable Development and several National Research Council committees.

Jonathan Overpeck is both a Professor and Director of the Institute for the Study of Planet Earth at the University of Arizona. Dr. Overpeck's research focuses on global change dynamics. In particular, his research aims to reconstruct and understand the full range of climate system variability, recognize and anticipate possible "surprise" behavior in the climate system, understand how the earth system responds to changes in climate forcing, and detect and attribute environmental change to various natural (e.g., volcanic, solar) and non-natural (e.g., greenhouse gases or tropospheric aerosol) forcing mechanisms. A major research component is to build an understanding of how key tropical systems vary on timescales longer than seasons and years. This work is motivated by hints that the key tropical climate systems have surprising modes of variability not present in the short instrumental record, and that shifts between modes can take place over intervals as short as a few years. Dr. Overpeck served on the National Research Council's U.S. National Committee for the International Union for Quaternary Research.

Dorothy Peteet is both a Senior Research Scientist at the NASA Goddard Institute for Space Science and an Adjunct Senior Research Scientist at the Lamont Doherty Earth Observatory of Columbia University. Dr. Peteet's research includes both climate modeling and paleoclimatic investigation using pollen/macrofossil studies. She has documented terrestrial abrupt climate changes and then used general circulation models (GCMs) to explore the sensitivity of the models to various rapid forcings, including changes in ocean temperature. Her research focuses on understanding the mechanisms and causes of rapid climate change.

Roger Pielke, Jr. is a scientist at the National Center for Atmospheric Research. Dr. Pielke focuses on the use of scientific research in the decision-making processes of public and private individuals and groups. In particular, his areas of research interest are weather impacts on society, global climate change policy, and science policy. Dr. Pielke is a board member of the Board on Atmospheric Sciences and Climate and has served on several National Research Council committees.

Ray Pierrehumbert is a Professor of Geophysical Sciences at the University of Chicago. His research is generally concerned with how climate works as a system. A recurrent theme to his research is the determination of the earth's relative humidity distribution, which is the key to many climate change problems. Dr. Pierrehumbert also maintains research interests in geophysical fluid dynamics, particularly as related to baroclinic instability, storm track structure, and planetary wave propagation.

Peter Rhines is a Professor of Oceanography and Atmospheric Sciences at the University of Washington. Dr. Rhines's research involves the theory of general ocean circulation, including ocean waves and eddies. His research also involves investigation of atmosphere and climate dynamics, particularly in the subpolar oceans. Dr. Rhines has an active seagoing program in the Labrador Sea, studying climate change and the physics of deep convection. Professor Rhines is a member of the National Academy of Sciences, and Fellow of the American Geophysical Union and the American Meteorological Society. He has served on numerous National Research Council committees.

Thomas Stocker is a Professor of Climate and Environmental Physics at the Physics Institute of the University of Bern, Switzerland. His research includes studies on the dynamics of the climate system, climate modeling, past and future climate change, and abrupt climate change. Current projects include the development of low-order coupled climate models, investigations on climate change and biogeochemical cycles, and studies documenting CO_2 variations during abrupt climate change. He was a Coordinating Lead Author of the chapter "Physical climate processes and feedbacks" in "Climate Change 2001: The Scientific Basis," a Contribution of Working Group I to the Third Assessment Report of the Intergovernmental Panel on Climate Change.

Lynne D. Talley is a Professor at the Scripps Institution of Oceanography. Her research is diverse, ranging from the formation, circulation, and distribution of water masses to current transport velocities. She is a member of the NSF Geosciences Advisory Council, the International WOCE Science

Steering Group and the NASA Earth System Science Advisory Committee. She is an editor of the *Journal of Physical Oceanography.* She has served on several National Research Council committees and panels including the Climate Research Committee and the Global-Ocean-Atmosphere-Land System Panel.

John M. Wallace is a Professor of Atmospheric Sciences at the University of Washington. His research has improved our understanding of global climate and its year-to-year and decade-to-decade variations, through the use of observational data. He has been instrumental in identifying and understanding a number of atmospheric phenomena, such as the spatial patterns in month-to-month and year-to-year climate variability, including the one through which the El Niño phenomenon in the tropical Pacific influences climate over North America. Dr. Wallace is a member of the National Academy of Sciences and has chaired several National Research Council panels including the Panel on Reconciling Temperature Observations, the Panel on Dynamic Extended Range Forecasting, and the Advisory Panel for the Tropical Ocean/Global Atmosphere (TOGA).

Staff

Alexandra Isern was a Program Officer with the Ocean Studies Board when this study began and served as Study Director for the activity. She received her Ph.D. in Marine Geology from the Swiss Federal Institute of Technology in 1993. Dr. Isern was a lecturer in Oceanography and Geology at the University of Sydney, Australia from 1994-1999. Her research focuses on the influences of paleoclimate and sea level variability on ancient reefs. Dr. Isern was co-chief scientist for Ocean Drilling Program Leg 194 that investigated the magnitudes of ancient sea level change. In July of 2001, Dr. Isern became a Program Director with the National Science Foundation.

John Dandelski is a Research Associate with the Ocean Studies Board and received his M.A. in Marine Affairs and Policy from the Rosenstiel School of Marine and Atmospheric Science, University of Miami. His research focused on commercial fisheries' impacts to the benthic communities of Biscayne Bay. As a graduate research intern at the Congressional Research Service he worked on fisheries and ocean health issues. Mr. Dandelski served as the RSMAS Assistant Diving Safety Officer and was involved in fisheries, coral, underwater archaeology, and ocean exploration projects.

Chris Elfring is Director of the NRC's Polar Research Board, where she is responsible for all aspects of the Board's strategic planning, project development and oversight, financial management, and personnel. Since joining the Polar Research Board in 1996, Ms. Elfring has overseen or directed studies such as *Enhancing NASA's Contributions to Polar Science* (2001), *The Gulf Ecosystem Monitoring Program: First Steps Toward a Long-term Research Plan* (2001), *Future Directions for NSF's Arctic Natural Sciences Program* (1998), and *The Bering Sea Ecosystem* (1996).

Morgan Gopnik is Director of the NRC's Ocean Studies Board. She earned a B.Sc. in Physical Geography and Environmental Studies from McGill University in Montreal, Canada, and a M.S. in Environmental Engineering Science from the California Institute of Technology (Caltech) in Pasadena, California, where she conducted research in fluid dynamics. Since Ms. Gopnik assumed leadership at the Ocean Studies Board in 1996, the board has produced over 30 reports providing independent, authoritative, objective advice to government agencies and the public about all aspects of ocean science and policy

Megan Kelly received her B.S. in Marine Science from the University of South Carolina in May 1999. She was a Senior Project Assistant for the Ocean Studies Board until April 2001 when she joined the Information and Technology Services division of The National Academies.

Jodi Bachim is a Senior Project Assistant for the Ocean Studies Board. She received her B.S. in zoology from the University of Wisconsin-Madison in 1998. Since starting with the Board in May 1999, Ms. Bachim has worked on several studies regarding fisheries, geology, nutrient over-enrichment, and marine mammals. In January 2002, she started taking classes for her M.S. in Environmental Biology.

Ann Carlisle is an Administrative Associate for the Polar Research Board. She received her B.A. in sociology from George Mason University in 1997. Ms. Carlisle, who was formerly with the Ocean Studies Board, has worked on studies regarding marine protected areas, fisheries stock assessments, polar geophysical data sets, monitoring plans in the Gulf of Alaska, and other topics.

B

Board Rosters

This study was a joint activity of the Ocean Studies Board, Polar Research Board, and Board on Atmospheric Sciences and Climate. The three boards, whose volunteer members are listed below, worked together to plan the study and provide substantive and administrative oversight throughout the study process.

OCEAN STUDIES BOARD

KENNETH BRINK (Chair), Woods Hole Oceanographic Institution, Woods Hole, Massachusetts

ARTHUR BAGGEROER, Massachusetts Institute of Technology, Cambridge Massachusetts

JAMES COLEMAN, Louisiana State University Baton Rouge, Louisiana

CORTIS K. COOPER, Chevron Petroleum Technology Company, San Ramon, California

LARRY CROWDER, Duke University Marine Laboratory, Beaufort, North Carolina

G. BRENT DALRYMPLE, Oregon State University, Corvallis, Oregon

EARL H. DOYLE, Shell Oil (ret.), Sugarland, Texas

ROBERT DUCE, Texas A & M University, College Station, Texas

D. JAY GRIMES, University of Southern Mississippi, Ocean Springs, Mississippi

RAY HILBORN, University of Washington, Seattle, Washington

MIRIAM KASTNER, Scripps Institution of Oceanography, La Jolla, California
CINDY LEE, State University of New York, Stony Brook, New York
ROGER LUKAS, University of Hawaii, Manoa, Hawaii
BONNIE MCCAY, Rutgers University, Cook College, New Brunswick, New Jersey
RAM MOHAN, Blasland, Bouck & Lee, Inc., Annapolis, Maryland
SCOTT NIXON, University of Rhode Island, Narragansett, Rhode Island
NANCY RABALAIS, Louisiana Universities Marine Consortium, Chauvin, Louisiana
WALTER SCHMIDT, Florida Geological Survey, Tallahassee, Florida
JON G. SUTINEN, University of Rhode Island, Kingston, Rhode Island
NANCY TARGETT, University of Delaware, Lewes, Delaware
PAUL TOBIN, Xtria, Chantilly, Virginia

NRC Staff

MORGAN GOPNIK, Director
SUSAN ROBERTS, Senior Program Officer
DAN WALKER, Senior Program Officer
JOANNE C. BINTZ, Program Officer
JENNIFER MERRILL, Program Officer
TERRY SCHAEFER, Program Officer
JOHN DANDELSKI, Research Associate
ROBIN MORRIS, Financial Officer
SHIREL SMITH, Office Manager
JODI BACHIM, Senior Project Assistant
NANCY CAPUTO, Senior Project Assistant
DENISE GREENE, Senior Project Assistant
DARLA KOENIG, Senior Project Assistant
JULIE PULLEY, Project Assistant

POLAR RESEARCH BOARD

DONAL T. MANAHAN, *Chair*, University of Southern California, Los Angeles
RICHARD B. ALLEY, Pennsylvania State University, University Park
ROBIN BELL, Lamont-Doherty Earth Observatory, Palisades, New York
AKHIL DATTA-GUPTA, Texas A&M University, College Station

HENRY P. HUNTINGTON, Huntington Consulting, Eagle River,
Alaska
AMANDA LYNCH, University of Colorado, Boulder
ROBIE MACDONALD, Fisheries and Oceans Canada, British Columbia
MILES MCPHEE, McPhee Research Company, Naches, Washington
P. BUFORD PRICE, JR., University of California, Berkeley
CAROLE L. SEYFRIT, Old Dominion University, Norfolk, Virginia
MARILYN D. WALKER, University of Alaska, Fairbanks

Ex-Officio Members

MAHLON C. KENNICUTT, Texas A&M University, College Station
(ex officio)
ROBERT RUTFORD, University of Texas, Dallas
PATRICK WEBBER, Michigan State University, East Lansing (ex officio)

NRC Staff

CHRIS ELFRING, Director
ANN CARLISLE, Senior Project Assistant

BOARD ON ATMOSPHERIC SCIENCES AND CLIMATE

ERIC J. BARRON (*Chair*), Pennsylvania State University, University
Park, Pennsylvania
SUSAN K. AVERY, Cooperative Institute for Research in Environmental
Sciences, University of Colorado, Boulder, Colorado
RAYMOND J. BAN, The Weather Channel, Inc., Atlanta, Georgia
HOWARD B. BLUESTEIN, University of Oklahoma, Norman,
Oklahoma
STEVEN F. CLIFFORD, Cooperative Institute for Research in
Environmental Sciences, University of Colorado, Boulder, Colorado
GEORGE L. FREDERICK, Vaisala Meteorological Systems, Inc.,
Boulder, Colorado
JUDITH L. LEAN, Naval Research Laboratory, Washington, D.C.
MARGARET A. LEMONE, National Center for Atmospheric Research,
Boulder, Colorado
MARIO J. MOLINA, Massachusetts Institute of Technology, Cambridge,
Massachusetts

ROGER A. PIELKE, JR., Cooperative Institute for Research in
  Environmental Sciences, University of Colorado, Boulder, Colorado
MICHAEL J. PRATHER, University of California, Irvine, California
WILLIAM J. RANDEL, National Center for Atmospheric Research,
  Boulder, Colorado
ROBERT T. RYAN, WRC-TV, Washington, D.C.
THOMAS F. TASCIONE, Sterling Software, Inc., Bellevue, Nebraska
ROBERT A. WELLER, Woods Hole Oceanographic Institution, Woods
  Hole, Massachusetts
ERIC F. WOOD, Princeton University, Princeton, New Jersey

Ex Officio Members

DARA ENTEKHABI, Massachusetts Institute of Technology, Cambridge,
  Massachusetts
EUGENE M. RASMUSSON, University of Maryland, College Park,
  Maryland
PAUL L. SMITH, South Dakota School of Mines and Technology, Rapid
  City, South Dakota

NRC Staff

ELBERT W. (JOE) FRIDAY, JR., Director
LAURIE S. GELLER, Senior Program Officer
PETER A. SCHULTZ, Senior Program Officer
VAUGHAN C. TUREKIAN, Program Officer
DIANE GUSTAFSON, Administrative Associate
ROBIN MORRIS, Financial Officer
TENECIA A. BROWN, Project Assistant
CARTER W. FORD, Project Assistant

C

Workshop Agenda

Abrupt Climate Change: Science & Public Policy Workshop
October 30-31, 2000
Lamont-Doherty Earth Observatory

Monday, October 30ᵗʰ

7:30 a.m.	Registration	
8:00 a.m.	Continental Breakfast	
8:30 a.m.	Welcome and Introduction	Richard Alley
8:45 a.m.	Keynote address	Wallace Broecker
9:30 a.m.	Evidence of abrupt climate change	Jeff Severinghaus
10:00 a.m.	Discussion	Jean Lynch-Stieglitz
10:30 a.m.	Break	
10:45 a.m.	Mechanisms of abrupt climate change	Mark Cane
11:15 a.m.	Discussion	Issac Held
11:45 a.m.	Working Lunch	
12:45 p.m.	Impacts of abrupt climate change	David Bradford
1:15 p.m.	Discussion	Gary Yohe
1:45 p.m.	Introduction of charge for Breakout group 1	Richard Alley
2:00 p.m.	Breakout group 1	
4:00 p.m.	Wrap-up of Breakout group 1	
5:00 p.m.	Adjourn for the day	
5:15 p.m.	Reception–Lamont Hall	

Tuesday, October 31ˢᵗ

8:00 a.m.	Continental Breakfast	
8:30 a.m.	Introduction to the day's events	Richard Alley
9:00 a.m.	Ongoing observations	Bob Dickson
9:30 a.m.	Discussion	Peter Schlosser
10:00 a.m.	Simulations of abrupt climate change	Sylvie Joussaume
10:30 a.m.	Discussion	John Kutzbach
11:00 a.m.	Break	
11:15 a.m.	Atmospheric dynamics	Grant Branstator
11:45 a.m.	Discussion	Tony Broccoli
12:15 p.m.	Working Lunch	
1:15 p.m.	Policy advice	William Ascher
1:45 p.m.	Discussion	Karen Smoyer
2:15 p.m.	Introduction to Breakout group 2	Richard Alley
2:30 p.m.	Breakout group 2	
4:00 p.m.	Wrap-up of Breakout group 2	
5:00 p.m.	Closing remarks	Richard Alley
5:30 p.m.	Meeting Adjourns	

D

Workshop Participants

Abrupt Climate Change: Science and Public Policy Workshop
October 30-31, 2000
Lamont-Doherty Earth Observatory

Richard Alley, *Chair, Pennsylvania State University*
Bob Anderson, *Lamont-Doherty Earth Observatory*
David Anderson, *National Oceanic and Atmospheric Administration*
William Ascher, *Claremont McKenna College*
Gerard Bond, *Lamont-Doherty Earth Observatory*
David Bradford, *Princeton University*
Grant Branstator, *National Center for Atmospheric Research*
Wallace Broecker, *Lamont-Doherty Earth Observatory*
Tony Broccoli, *Geophysical Fluid Dynamics Laboratory*
Mark Cane, *Lamont-Doherty Earth Observatory*
Amy Clement, *Lamont-Doherty Earth Observatory*
Edward Cook, *Lamont-Doherty Earth Observatory*
Peter de Menocal, *Lamont-Doherty Earth Observatory*
George Denton, *University of Maine*
Bob Dickson, *Centre for Environment, Fisheries and Aquaculture Science*
Mary Elliot, *Lamont-Doherty Earth Observatory*
Christa Farmer, *Lamont-Doherty Earth Observatory*
Peter Gent, *National Center for Atmospheric Research*
Morgan Gopnik, *National Research Council, Ocean Studies Board*
Isaac Held, *Geophysical Fluid Dynamics Laboratory*
Terry Hughes, *University of Maine*

Alexandra Isern, *National Research Council, Ocean Studies Board*
Stan Jacobs, *Lamont-Doherty Earth Observatory*
Sylvie Joussaume, *Laboratoire des Sciences du Climat et de l'Environnement*
Lloyd Keigwin, *Woods Hole Oceanographic Institution*
Klaus Keller, *Princeton University*
Megan Kelly, *National Research Council, Ocean Studies Board*
John Kermond, *National Oceanic and Atmospheric Administration*
Athanasios Koutavas, *Lamont-Doherty Earth Observatory*
Yochanan Kushnir, *Lamont-Doherty Earth Observatory*
John Kutzbach, *University of Wisconsin*
Scott Lehman, *University of Colorado*
Jean Lynch-Stieglitz, *Lamont-Doherty Earth Observatory*
Tom Marchitto, *Lamont-Doherty Earth Observatory*
Jochem Marotzke, *University of Southampton*
Doug Martinson, *Lamont-Doherty Earth Observatory*
Bruce Molnia, *U.S. Geological Survey*
Ken Mooney, *National Oceanic and Atmospheric Administration*
Pierre Morel, *University of Maryland, Baltimore County*
William Nordhaus, *Yale University*
Jonathan Overpeck, *University of Arizona*
Dorothy Peteet, *NASA Goddard Institute for Space Studies; Lamont-Doherty Earth Observatory*
Raymond Pierrehumbert, *University of Chicago*
Peter Rhines, *University of Washington*
Peter Schlosser, *Lamont-Doherty Earth Observatory*
Gavin Schmidt, *National Aeronautics and Space Administration*
Jeff Severinghaus, *Scripps Institution of Oceanography*
John Shepherd, *University of Southampton*
Karen Smoyer, *University of Alberta*
David Stahle, *University of Arkansas*
Thomas Stocker, *University of Bern*
Lynne Talley, *Scripps Institution of Oceanography*
Duane Thresher, *National Aeronautics and Space Administration*
James Todd, *National Oceanic and Atmospheric Administration*
Sushel Unninayar, *National Aeronautics and Space Administration*
Dirk Verschuren, *University of Ghent*
Martin Visbeck, *Lamont-Doherty Earth Observatory*
Sean Willard, *National Oceanic and Atmospheric Administration*
Gary Yohe, *Wesleyan University*

E

Impacts Workshop Program

Workshop on Economic and Ecological Impacts of Abrupt Climate Change
March 22-23, 2001
Foundry Building, Washington, D.C.

Thursday, March 22nd

9:00 AM	Briefing of participants on background on abrupt climate change (Peteet, Nordhaus)
10:00 AM	First session on ecosystem impacts (with break)

A. *Vegetational change*
Allen on forest/woodland shifts; Cook on forest response; Koenig on seed production; Swetnam on fire & ecosystems

B. *Animals & Climate Change*
Leopold on mid-US phenology; Daszak on disease;

12:30 PM	LUNCH
1:30 PM	First session on economic impacts (with break)

A. *Methods*
Nordhaus on alternative approaches; Smith on IPCC report

B. *Unmanaged or unmanageable systems*
Reilly on agriculture; Yohe on oceans; Mendelsohn on forests; Ausubel on fisheries

4:00 PM   General Discussion
5:00 PM   ADJOURN FOR THE DAY
6:00 PM   (Optional) The National Academies, Polar Research Board
              Public Lecture: *Climate Change: From the Poles to the
              World*, presented by Dr. Richard Alley. The Cecil and Ida
              Green Building, Room 104, 2001 Wisconsin Avenue, NW.

Friday, March 23rd

9:00 AM   Second session on ecosystem impacts
              Hydrological cycle
              Inouye on snowfall and altitudinal migrants; Kling on
              freshwater ecosystems; Lowell on Arctic and glacial
              response; Dyurgerov on alpine glaciers
11:00 AM   Second session on economic impacts
              Thermohaline circulation studies
              Toth, Tol, and Keller on thermohaline circulation reversal
12:00 PM   LUNCH
1:00 PM   Third session on societal impacts and responses
              A. *Human and societal responses*
                  Weiss on ancient civilizations
              B. *Responses in today's world*
Weyant on technology; Pielke on adaptation
2:20 PM   General conclusions from the different areas
3:15 PM   Tentative recommendations for the report
4:00 PM   ADJOURN

F

Impacts Workshop Participants

Workshop on Economic and Ecological Impacts of Abrupt Climate Change
March 22-23, 2001
Foundry Building, Washington, D.C.

Craig Allen, *U.S. Geological Survey*
Jesse Ausubel, *The Rockefeller University*
Edward Cook, *Columbia University*
Peter Daszak, *University of Georgia*
Mark Dyurgerov, *University of Colorado*
David Inouye, *University of Maryland*
Alexandra Isern, *National Research Council*
Klaus Keller, *Princeton University*
Megan Kelly, *National Research Council*
George Kling, *University of Michigan*
Walter Koenig, *University of California, Berkeley*
Carl Leopold, *Cornell University*
Thomas Lowell, *University of Cincinnati*
Robert Mendelsohn, *Yale University*
William Nordhaus, *Yale University*
Dorothy Peteet, *Goddard Institute for Space Studies*
Roger Pielke, Jr., *National Center for Atmospheric Research*
John Reilly, *Massachusetts Institute for Technology*
Joel Smith, *Stratus Consulting, Inc.*
Thomas Swetnam, *University of Arizona*

Richard Tol, *Hamburg University*
Ferenc Toth, *Potsdam Institute for Climate Impact Research*
Harvey Weiss, *Yale University*
John Weyant, *Stanford University*
Gary Yohe, *Wesleyan University*

6

Acronym List

AO	Arctic Oscillation
AAO	Antarctic Oscillation
CFCs	chloroflourocarbons
DSOW	Denmark Strait Overflow Water
ENSO	El Niño/Southern Oscillation
GCM	general circulation models
GFDL	Geophysical Fluid Dynamics Laboratory
GISP2	Greenland Ice Sheet Project 2
IGBP	International Geosphere-Biosphere Programme
ITCZ	Intertropical Convergence Zones
LPTM	Late Paleocene Thermal Maximum
LSW	Labrador Sea Water
NAO	North Atlantic Oscillation
NBER	National Bureau of Economic Research
NEADW	Northeast Atlantic Deep Water

PAGES	PAst Global changES program
PDO	Pacific Decadal Oscillation
PETM	Paleocene-Eocene Thermal Maximum
PNA	Pacific North American pattern
SST	Sea surface temperature
TAO	Tropical Atmosphere-Ocean
THC	thermohaline circulation
THCC	thermohaline circulation collapse
USGCRP	US Global Change Research Program
WOCE	World Ocean Circulation Experiment
YD	Younger Dryas

Index

A

Aerosols, 83-84, 86, 138
  chlorofluorocarbons, 47, 66, 165
  dust, 21, 25, 27, 30, 36, 80, 83, 86, 144
Africa
  disease, 147
  drought, 55, 74
  Eemian climate change, 39
  Younger Dryas, 33-34, 35
Agriculture, 17, 53, 57, 114, 119, 123-125,
    128, 133, 137, 138, 141-142,
    143, 145, 148, 152
  see also Drought
Albedo, 75, 78, 79, 86
Animal and Plant Health Inspection Service,
    158
Annular variation, *see* Interannual
    variations
Antarctica, 36, 65-67, 69, 77, 78, 156, 162
  Younger Dryas, 26, 35
Antarctic Oscillation, 49
Arctic Ocean, 65, 114, 117, 140
Arctic Oscillation (AO), 48, 49, 50, 52, 57-
    58, 60, 62, 65, 67-68, 113
Atlantic Ocean, 36-37, 38, 60-69, 71, 72,
    76-77, 80-81, 82, 84, 162
  Eemian climate change, 39
  El Niño, impacts on, 50, 52, 110
  Holocene rapid climate change, 41

North Atlantic Oscillation (NAO), 44,
    49, 57-58, 60-61, 62, 65, 67-68,
    90, 113, 134
  thermohaline circulation, 77, 80-81, 82,
    84-85, 90, 96, 102, 103, 108-
    117, 156
  tropical variability, 48, 52-53, 62, 78,
    81-82, 110
  Younger Dryas, 34, 35-36, 76-77
Atmosphere-ocean interactions, 47, 62, 67-
    69, 71, 75-76, 80-83, 84-85, 90,
    108, 110, 112
  see also El Niño/Southern Oscillation
Antarctic Oscillation, 49
Arctic Oscillation (AO), 48, 49, 50, 52,
    57-58, 60, 62, 65, 67-68, 113
  committee research recommendations, 2,
    154
  modeling of, 75-76, 82-83, 94, 101,
    114, 154
    general circulation models (GCMs),
      83, 92, 97, 99-100, 101, 103,
      104
  North Atlantic Oscillation (NAO), 44,
    49, 57-58, 60-61, 62, 65, 67-68,
    90, 113, 134
  sea surface processes, 47, 50, 51, 79, 84
Atmospheric processes, other, 71, 73, 80-86
  see also Drought; Greenhouse gases;
    Precipitation; Wind

223

M

N

O